T0328110

Long-Term Farming Systems Research

Ensuring Food Security in Changing Scenarios

Long-Term Farming Systems Research

Ensuring Food Security in Changing Scenarios

Edited by

Gurbir S. Bhullar

Department of International Cooperation
Research Institute of Organic Agriculture (FiBL)
Frick, Switzerland

Amritbir Riar

Department of International Cooperation
Research Institute of Organic Agriculture (FiBL)
Frick, Switzerland

Academic Press is an imprint of Elsevier
125 London Wall, London EC2Y 5AS, United Kingdom
525 B Street, Suite 1650, San Diego, CA 92101, United States
50 Hampshire Street, 5th Floor, Cambridge, MA 02139, United States
The Boulevard, Langford Lane, Kidlington, Oxford OX5 1GB, United Kingdom

Copyright © 2020 Elsevier Inc. All rights reserved.

No part of this publication may be reproduced or transmitted in any form or by any means, electronic or mechanical, including photocopying, recording, or any information storage and retrieval system, without permission in writing from the publisher. Details on how to seek permission, further information about the Publisher's permissions policies and our arrangements with organizations such as the Copyright Clearance Center and the Copyright Licensing Agency, can be found at our website: www.elsevier.com/permissions.

This book and the individual contributions contained in it are protected under copyright by the Publisher (other than as may be noted herein).

Notices

Knowledge and best practice in this field are constantly changing. As new research and experience broaden our understanding, changes in research methods, professional practices, or medical treatment may become necessary.

Practitioners and researchers must always rely on their own experience and knowledge in evaluating and using any information, methods, compounds, or experiments described herein. In using such information or methods they should be mindful of their own safety and the safety of others, including parties for whom they have a professional responsibility.

To the fullest extent of the law, neither the Publisher nor the authors, contributors, or editors, assume any liability for any injury and/or damage to persons or property as a matter of products liability, negligence or otherwise, or from any use or operation of any methods, products, instructions, or ideas contained in the material herein.

Library of Congress Cataloging-in-Publication Data
A catalog record for this book is available from the Library of Congress

British Library Cataloguing-in-Publication Data
A catalogue record for this book is available from the British Library

ISBN: 978-0-12-818186-7

For information on all Academic Press publications visit our website at https://www.elsevier.com/books-and-journals

Publisher: Charlotte Cockle
Acquisitions Editor: Nancy Maragioglio
Editorial Project Manager: Ruby Smith
Production Project Manager: Kiruthika Govindaraju
Cover Designer: Alan Studholme

Typeset by TNQ Technologies

To the wisdom of those who think ahead of time and leave a wealth of knowledge for future generations

Contents

SECTION III Maximizing outcomes through networking and capacity development

Contributors

Daniele Antichi
Department of Agriculture, Food and Environment, University of Pisa, Pisa, Italy

Keith Bamford
Department of Plant Science, University of Manitoba, Winnipeg, MB, Canada

Maya V. Belichenko
All-Russian Institute of Agrochemistry named after D. Pryanishnikov, Geographic Network of Field Experiments Department, Moscow, Russia; Leading Researcher, Geographical Network of Field Experiments, All-Russian Institution of Agrochemistry named after D. Pryanishnikov, Moscow, Russia

Gurbir S. Bhullar
Department of International Cooperation, Research Institute of Organic Agriculture (FiBL), Frick, Switzerland

Tim Brooker
Foundation for Arable Research, Christchurch, New Zealand

Stefano Canali
CREA-Research Centre Agriculture & environment, Rome, Italy

Michelle Carkner
Department of Plant Science, University of Manitoba, Winnipeg, MB, Canada

Danilo Ceccarelli
CREA-Research Centre for Olive, Citrus and Tree Fruit, Rome, Italy

Richard Chynoweth
Foundation for Arable Research, Christchurch, New Zealand

Corrado Ciaccia
CREA-Research Centre Agriculture & environment, Rome, Italy

Calvin Dick
Department of Plant Science, University of Manitoba, Winnipeg, MB, Canada

Martin H. Entz
Department of Plant Science, University of Manitoba, Winnipeg, MB, Canada

Andreas Fliessbach
Research Institute of Organic Agriculture (FiBL), Department of Soil Sciences, Frick, Switzerland

Margaret J. Glendining
Rothamsted Research, Computational and Analytical Sciences Department, Harpenden, Herts, United Kingdom

Meike Grosse
Leibniz Centre for Agricultural Landscape Research (ZALF), Müncheberg, Germany

Carsten Hoffmann
Leibniz Centre for Agricultural Landscape Research (ZALF), Müncheberg, Germany

Abie Horrocks
Foundation for Arable Research, Christchurch, New Zealand

Radosław Kaczyński
Institute of Soil Science and Plant Cultivation, State Research Institute, Pulawy, Poland

Tamanpreet Kaur
School of Organic Farming, Punjab Agricultural University, Ludhiana, Punjab, India

Hans-Martin Krause
Research Institute of Organic Agriculture (FiBL), Department of Soil Sciences, Frick, Switzerland

Jan Kuś
Institute of Soil Science and Plant Cultivation, State Research Institute, Pulawy, Poland

Artur Łopatka
Institute of Soil Science and Plant Cultivation, State Research Institute, Pulawy, Poland

Andy J. Macdonald
Rothamsted Research, Sustainable Agriculture Sciences Department, Harpenden, Herts, United Kingdom

Paul Mäder
Research Institute of Organic Agriculture (FiBL), Department of Soil Sciences, Frick, Switzerland

Mariusz Matyka
Institute of Soil Science and Plant Cultivation, State Research Institute, Pulawy, Poland

Jochen Mayer
Agroscope, Department Agroecology and Environment, Zurich, Switzerland

Wiesław Oleszek
Institute of Soil Science and Plant Cultivation, State Research Institute, Pulawy, Poland; Biochemistry and Crop Quality, Institute of Soil Science and Plant Cultivation, State Research Institute, Pulawy, Poland

Nick Poole
Foundation for Arable Research, Christchurch, New Zealand

Paul R. Poulton
Rothamsted Research, Sustainable Agriculture Sciences Department, Harpenden, Herts, United Kingdom

David S. Powlson
Rothamsted Research, Sustainable Agriculture Sciences Department, Harpenden, Herts, United Kingdom

Nick Pyke
Foundation for Arable Research, Christchurch, New Zealand

Amritbir Riar
Department of International Cooperation, Research Institute of Organic Agriculture (FiBL), Frick, Switzerland

Phil Rolston
Foundation for Arable Research, Christchurch, New Zealand

Vladimir A. Romanenkov
Lomonosov Moscow State University, Soil Science Faculty, Eurasian Center for Food Security, Moscow, Russia; All-Russian Institute of Agrochemistry named after D. Pryanishnikov, Geographic Network of Field Experiments Department, Moscow, Russia; Head of Department, Agrochemistry and Plant Bochemistry, Lomonosov Moscow State University, Moscow, Russia; Leading Researcher, Geographical Network of Field Experiments, All-Russian Institution of Agrochemistry named after D. Pryanishnikov, Moscow, Russia

Olga V. Rukhovich
All-Russian Institute of Agrochemistry named after D. Pryanishnikov, Geographic Network of Field Experiments Department, Moscow, Russia; Head of Department, Geographic Network of Field Experiments, All-Russian Institute of Agrochemistry named after D. Pryanishnikov, Moscow, Russia

K. Scow
Professor, LAWR, University of California, Davis, CA, United States

Lyudmila K. Shevtsova
All-Russian Institute of Agrochemistry named after D. Pryanishnikov, Geographic Network of Field Experiments Department, Moscow, Russia; Principal Researcher, Geographic Network of Field Experiments Department, All-Russian Institute of Agrochemistry named after D. Pryanishnikov, Moscow, Russia

Grzegorz Siebielec
Institute of Soil Science and Plant Cultivation, State Research Institute, Pulawy, Poland

Xenia Specka
Leibniz Centre for Agricultural Landscape Research (ZALF), Müncheberg, Germany

April Stainsby
Department of Plant Science, University of Manitoba, Winnipeg, MB, Canada

Katherine Stanley
Department of Plant Science, University of Manitoba, Winnipeg, MB, Canada

Nikolai Svoboda
Leibniz Centre for Agricultural Landscape Research (ZALF), Müncheberg, Germany

N. Tautges
Chief Scientist, Agricultural Sustainability Institute, Russell Ranch Sustainable Agriculture Facility, University of California Davis, Davis, CA, United States

Joanne Thiessen Martens
Department of Plant Science, University of Manitoba, Winnipeg, MB, Canada

S.S. Walia
School of Organic Farming, Punjab Agricultural University, Ludhiana, Punjab, India

Sarah Wilcott
Department of Plant Science, University of Manitoba, Winnipeg, MB, Canada

Foreword

In a world where the quantity of food is trumped over nutritional quality as expressed by its diversity and nutrient density—even as a substantial portion of that quantity ends up wasted—food security remains a challenge and major preoccupation of governments, international development agencies, and not-for-profit civil society organizations. Food security issues, occupying a large part of national and international development agendas, are clearly not being resolved within the present industrial and conventional agriculture and food systems. These systems are only adding to the climate crisis, not to mention the more recent pandemic, greatly facilitated by globalization which has weakened people's general health status from both air pollution and overexposure to endocrine disruptors used in the production system. The winter 2019–20 is so far the warmest since temperatures are being recorded. Our ice caps, north and south, are melting at a record speed, while droughts, fires, and floods are a constant reminder that we are already facing major impact of climate change on our food system and its basis, the soil. Too much has already been written about the challenges and too little done, even while the predicted changes and associated crises are upon us, creating havoc and turning many decision makers into "chickens running around without their heads," looking for quick fixes. Instead, the world community (and its decision makers) could rely on a science that has already developed solutions to not only adapt to the new reality but also mitigate further system disintegration.

As reported in this landmark book, scientists and farmers with foresight have planned decades ago for harsher times, as major catastrophes have impacted food production, and raised questions about our food and food production system's resilience in past times already. How to increase the system's resilience? How to be prepared for major climatic, but also in the shorter-term changing weather, patterns? It has been recognized by the pioneers of Long-Term Experiments in Agriculture (LTEs) that in addition to the short-term research in dealing with pests and diseases that affect our field crops (including vegetables and fruit trees), many other factors are important; the impact of fertilizers (natural and synthetic) on yields, selection of different varieties to adapt to ecological conditions, and good agronomic practices are all essential to assure good productivity over the long term. It was also realized that food security is not about a one-off high yield, but preferably a stable production over the years, with as little variation as possible, in other terms, resilient production, as well as to maintain a healthy market and prices.

It is this insight and concern that led to the LTEs in the first place, many of which are described in this book in great detail. It is also a call for one of the major difficulties of LTEs, the allocation of funding on a secure long-term basis. This is never easy, particularly in our present time when everything has to be run on projects, with an average lifetime of 3 years at best, with requirements for quickly publishable results in top journals. Stability is not easy to achieve when donors and investors from private or public funding agencies are demanding quick returns on their investments.

Other pressure on LTEs comes from urban developments on the periphery infringing at best and eliminating their bases to make room for urbanization, which is progressing relentlessly into agricultural land, and often into the experimental plots, of research stations that used to be in rural areas, now being surrounded and swallowed by human and industrial development.

The question therefore is on how to make the LTEs relevant to potential funders. Ideally these should be governments and their agricultural research institutions, who have the prime responsibility to provide farmers with the needed science and skills to manage their farms, under new and challenging situations, and fulfill their responsibility for food security. The regional R&D as well as international bodies would also have part of this responsibility. But it may be much more difficult, if not impossible to have a guaranteed long-term financial support, the way a national government could have a budget line item, called LTE in its Ministry of Agriculture budget. The trend over the past half century has been to transfer the fiscal responsibility to the private sector and nongovernmental funding sources. The lack of a secure governmental funding source is the biggest threat to the ongoing LTEs as well as the barrier to new ones. This trend needs to be reversed; food security, a human right, is a government matter, not private sector, and so the responsibility for it needs to reside with governments.

This is the time to seriously tackle the transformation of agriculture toward agroecological practices. Agroecology is often criticized that it may not be able to nourish the world, ignoring the reality that the present globalized commodity-based food system leaves 800 million hungry, while 1.5 billion are obese and diabetic. It would be of great help to have more solid data in the years ahead, given the expected changes in our climate, environmental, and socioeconomic situation, which will undoubtedly reflect back on who grow our food, how we grow our food, and where we grow our food. There is a compelling need for LTEs to guide, in a dynamic way, the transformation to truly sustainable production systems.

Given the fast changes of both environmental as well as economic conditions and availability of innovations that are affecting farmers and the way they grow food, LTEs must be dynamic and be responsive to the changes. These are often going in diverging directions, as seen in the past and now. On the one hand, we have an expansion of the industrial, high input model, with large monocrops, heavily mechanized, and automated systems from precision agriculture to the genetic modified crops. On the other hand, there are other agroecological practices, centered around people and the environment, with an eye on sustainability, diversity, social, and environmental responsibility and anchored in solid science. These are the type of comparisons, both in a very dynamic process, that need to be available, in all dimensions, well beyond pure yields as the yardstick for success. We are not there yet with what has been done and ongoing, but the authors in this book make the direction clear. It can only be hoped that the pressure from the industrial and conventional model lobby will not outweigh the one from the "heirs of the IAASTD" when it comes to allocation of government funding to the LTEs of the future, as well as the ongoing ones.

Thanks to the scientists and decision makers that supported LTEs in the wake of the elimination of a system approach to farming, we have a good basis to work from. There exists, within LTEs, potential to contribute the research toward transforming agricultural practices that will deliver on true sustainability. I recall how the Farming System Program at the International Institute of Tropical Agriculture, where I worked for 16 years, was abolished to favor economics and breeding programs in the mid-80s, when the reductionism, hallmark of the green revolution, took off and system thinking and approaches were deemed too complicated, long term, and lacking immediate measurable impact. The results of this approach are in plain sight now, with hugely degraded soils, dependence on agrochemicals, loss of agro-biodiversity, desertion of rural areas, continued poverty, hunger and increased environmental degradation, exemplified in the overshooting of the planetary boundaries.

Another path is possible, with LTEs providing the roadmap. With the emphasis of agriculture becoming part of the climate change solution, away from being part of the problem, this book will provide important evidence and gives guidance for transformative approaches in terms of resilience, adaptation, and mitigation and the quest for carbon sequestration. With examples from around the globe, the book also covers in great details the different approaches to LTEs. This book is a recognition of and appreciation for the scientists that stuck with the system's approach and the LTEs for their wisdom and endurance.

Hans R. Herren
President Millennium Institute, USA
World Food Prize (1995) Laureate

Preface

Agriculture, as basis of human civilization, exerts significant influence on several aspects of life on earth and also gets influenced by large majority of other human activities and decisions either directly or indirectly. As governments and development agencies across globe are pursuing to achieve the Global Sustainable Development Goals (SDGs) laid out by the United Nations, understanding the interactions of agricultural production with ecological, economic, and societal aspects is of unprecedented importance. Long-Term Experiments (LTEs) serve as an important tool in this regard. LTEs are not only great platforms for studying and extrapolating the effect of environmental, technological, and management regimes on various aspects of land productivity but also they generate knowledge on interactions of agriculture with biophysical environment. The unique value of LTEs in agriculture is highlighted by the fact that some of the important information goes undetected or is simply not possible to obtain in short-term studies. For instance, changes in soil carbon stocks and belowground biodiversity have important implications for the sustainability of farming systems, but these parameters can only be assessed in systematic studies encompassing over a few decades.

Starting with the classical experiments initiated in 1843 at Rothamsted, UK, there are a number of LTEs set up with different objectives in different agroecological zones of the world. These trials, on the one hand, are yielding a wealth of information every day, and on the other hand they face their unique challenges ranging from complex statistical analysis, maintenance of contextual credibility, and financial and institutional uncertainties. Some of these trials have successfully chiseled through several exceptional circumstances such as natural calamities, pandemics, and wars. It is valuable to collect the evidences and experiences from these trials to derive inferences for global agricultural systems as well as to cross-learn lessons for effective management of other/future LTEs. Scientific data generated from these LTEs are mostly published in the form of peer-reviewed publications or technical reports, but the limits of the scientific publication criteria mean that often a great wealth of knowledge produced in LTEs and the lessons learned by navigating through exceptional circumstances remain only with the custodians of the specific trials.

Thanks to the excellent panel of our contributing authors, in this book, we have attempted to offer a widest possible thematic and geographical coverage on LTEs. Experts from different institutions leading LTEs across globe have provided their respective perspectives on different aspects of LTEs, not only highlighting the unique knowledge contribution of LTEs but also discussing the unique challenges of effectively managing LTEs and maintaining their relevance to changing scenarios. Although the scientific findings from LTEs are of primary interest to agricultural and environmental scientists as well as policymakers across the globe, they are equally interesting for development agencies, NGOs, funding bodies, and extensionists. Sufficient detail on each topic is provided to address the interest of technically advanced readership, yet we have kept the language and expression as simple as possible for easy understanding of a nonexpert educated reader. We hope that this book will offer something for everyone interested in history, present, and future of our agroecosystem.

- The Editors

SECTION I

From the Editor's desk

CHAPTER

Long-term experiments in agriculture: stages, challenges, and precautions

Amritbir Riar, Gurbir S. Bhullar

Department of International Cooperation, Research Institute of Organic Agriculture (FiBL), Frick, Switzerland

Introduction

Since the dawn of domestication, humans are in a continuous quest to learn about phenomenal inputs from nature to produce, store, and process food according to their needs. Careful observation of nature led to a better understanding of natural processes responsible for plant growth, enabling interventions to alter these processes in a more controlled manner. This in-turn gave birth to experimentation in agriculture. Gradually, it was evident that there are many factors influencing plant growth, and it is not always possible to transfer the success story of one season to another, making the addition of time dimension as an important tool. With increasing knowledge, scope of experimentation broadened to cover for the changes that were not possible to observe in short term, i.e., in one or a few cropping seasons. This in-turn contributed to conception of the long-term experiments (LTEs) in agriculture. Besides the time factor, contemporary societal and technological developments are the other important reflections to and of LTEs.

Majority of the LTEs of modern times, initiated during mid-late 19th century, were primarily designed to compare/demonstrate crop response to specific inputs. With the improving understanding of complexity of biophysical and biogeochemical processes, the LTEs have evolved from comparing the response of inputs, mainly nutrients via mineral fertilizers and organic manures to systems comparisons (e.g., Rothamsted classical experiments (Macdonald, 2020), DOK trial (Krause et al., 2020)). However, attributing a field trial as a "long-term trial" is prone to the relativity of time with the functions of different system components, e.g., a cropping season can accommodate many life cycles of microbial communities. Still, the functionality of a microbial community in-relation to production system could be limited in real time. In other words, the active period for a system function can be a point of time beyond which a service performs at its best. Field trials in arable agriculture can be defined as "long-term trials" when inception vision of trial is well beyond decades. Many field trials attributed as long-term trials are active at different

Long-Term Farming Systems Research. https://doi.org/10.1016/B978-0-12-818186-7.00001-1
Copyright © 2020 Elsevier Inc. All rights reserved.

ages around the globe. Although the minimum time duration to designate a trial as long term is not well marked, around 10–15 years is considered optimal duration to serve many purposes. A German repository of LTEs—Soil as a Sustainable Resource for the Bioeconomy (BonaRes)—considers a trial as LTE if it has a minimum time span of 20 years and a static design and focuses on a scientific issue in the context of soil and yield (Grosse et al., 2020). According to this definition, a total of 200 LTEs across Germany have been identified. Time scale is not the only main criterion to define a trial as LTE; the other main criterion is experimental design, mainly in terms of the continuity of the repetitive cycle of a set of specified practices over time. These practices can, e.g., be crop rotations, soil amendments, nutrient addition, or climatic variations to study/validate the influences over time. From this perspective of management, LTEs can broadly categorize into three main stages, i.e., static, semistatic, or dynamic. At a given point of time, an LTE can only seem fitting to one of these stages, but one should be mindful that these stages are sequentially interchangeable.

Stages of long-term trials

Static

A static LTE can be defined as the experiment that has not changed any component of factors under investigation since the beginning, e.g., treatments, input rates, or crop rotation. A static experiment can, for instance, be a valuable asset to assess the perception of society on land use and sustainable food production, e.g., Broadbalk experiment (static form: 1844 to 1960) (Ashcroft et al., 1990; Macdonald, 2020). Nevertheless, a static experiment can easily become the victim of changes at landscape level, and cost-effectiveness of the experiment may come under question. This limitation could perhaps be overcome, if the experiment has a vibrant factor "something of a sensational character" (Lawes, 1882; Johnston and Poulton, 2018).

Semistatic: catching up with the time

Semistatic experiments are the experiments where efforts are made to keep the originality of experiment, but some changes are made to keep the experiment either relevant to the changing farming or societal contexts or changes were imperative due to the change in initial objective. Broadbalk wheat experiment (Rothamsted, UK) started in 1844 is probably the oldest continuous scientific experiment in existence (See Section Broadbalk Continuous Wheat experiment in Chapter 1). However, some changes were introduced after 116 years to keep the relevance to current agricultural practice (Macdonald, 2020). Another example is of the DOK trial (Switzerland), which is a systems comparison experiment, originally conceptualized as a semistatic experiment (for details, see Chapter 2). DOK trial respects the standards of label organization for inputs to biodynamic and bioorganic treatments, and equivalents are applied for the conventional treatments; thus, changes are required if label regulation changes, whereas two other systems comparison trials of Research Institute of Organic Agriculture (FiBL), namely, SysCom trials in India and Kenya, were conceptualized other way around (Forster et al., 2013; Musyoka et al., 2017). There were no local organic labels or standards available, but recommendations for conventional treatments with mineral inputs were available from national bodies in respective countries.

Dynamic

The dynamic long-term trials are the trials that are open to adapt to the changes in real time. The trials can be mimicking the farming practices of a particular region. The inception idea of these trials is often to evaluate the comparative performance of different farming practices under various production systems. Since these trials endeavor to stay relevant to changing farming practices, they bring value for local farming, e.g., by addressing challenges of current farming practices of the region. The main challenges faced by these trials include the difficulty in deriving inferences over the long term and scaling up results to landscape level because of the relatively frequent changes in management. Over time, these trials also tend to fall back to the semistatic state.

Challenges of long-term experiments

LTEs follow the milestones set in the past to measure the progress made over time. However, not all LTEs can be expected to last indefinitely. Ending an LTE can be the source of many unanswered questions, but sometimes reasons for concluding an LTE are more convincing than the ones for continuing it. It would be rational to shut down an LTE, particularly when it has lost its relevance to time or when the questions of inception are answered or are not relevant anymore, or when "something of a sensational character" (Lawes, 1882) is lost. The journey of an LTE is not only full of ups and downs, rather challenging at times for founders as well as for successors and supporters to keep alive the sensational charter without compromising the original aim. Here, we revisit some of the most critical challenges of LTEs.

Maintaining the relevance to current farming issues (time relevance)

One of the key issues facing LTEs around the globe is to stay relevant to fast-changing scenarios. Changes happen not only in the agricultural practices or land-use change perspective but also in societal preferences, e.g., shifting of food choices. These changes sometimes become imperative; otherwise, LTEs risk becoming the "monuments of their founders" (Johnston and Poulton, 2018).

One key point of change in LTEs revolves around management practices. Researchers often struggle with the questions of defining the "best possible practice" or reflecting the "current regional agricultural practice." Making changes to LTEs can be difficult and certainly provoke criticism at first place and a never-ending debate with ifs and buts. One of the main criticisms is often about undermining the previous research and difficulty in interpretation of data and results in the wake of changed practices. These conflicts often not only arise from significant changes, but sometimes the associated effects of a change are more concerning. For example, changes in crop varieties or crop rotation can stir the discussion not only on how the change affects the treatment but also on how will they contribute to changes in other factors (e.g., nutrient requirements of the crop rotation). Another primary debate revolves around the lack of a whole-system approach (Brooker et al., 2020) versus complexity of systems approach (Krause et al., 2020). In single or bifactorial agronomic trials (other than variety testing), genetic basis in different treatments is largely the same. However, in more complex systems comparison trials, the choice becomes more complicated. For instance, the crop variety best suited to a conventional system may not be the best fit for organic or other systems/practices; thus the argument of selection bias stays valid in both scenarios, i.e., same or best fit for each system. However, with increasing knowledge of biophysical interactions, understanding of and support for systems approach has grown in the recent years. However, defining system's boundaries and the differentiation of factors' effects remains to be a complex and challenging task. Moreover, farming and production systems do not stand still, and farmers are often more swift to respond to the emerging challenges like a new disease and pest or adapting to new technologies such as drip irrigation or crop varieties. Keeping up with these changes is challenging in LTEs; nevertheless, with appropriate feedback mechanisms, LTE management can try to reflect these changes within the allowed boundaries of LTE. For instance, in the example of SysCom India, an advisory committee comprising local organic and conventional farmers regularly provides their feedback to researchers, twice a year. Any changes to the LTE management should, however, to be standardized first, e.g., in separate field trials and only to be introduced to LTEs after having a thorough discussion with relevant stakeholders, e.g., farmers, researches, extensionists, and policymakers.

Trade-offs for adoption

While one aims to achieve system sustainability in the long as well as short term, for practical reasons, the interventions and their outcomes should also be relevant in the short term. As argued in the previous section, consistent management is necessary for extrapolation of results in space and time as well as for providing enough space to evolve the innovations. This can only be achieved through coexistence and coevolution of new technologies while considering the farmers' perspective. While making investment decisions, farmers are often faced with resource limitations, and chances are rare to invest at a point where the rate of return and resilience of system colimit each other. For example, "In a true no-till system, a farmer would not attempt to establish a ryegrass seed crop following a 12 t ha^{-1} wheat crop" (Brooker et al., 2020). Hence, the popularity of a system or treatment can be attributed to the economic gains with the ease of implementation or just a resistance toward change associated with no or minimum financial gain. However, it does not mean that popular systems will stay unchanged. Certain practices may not get popular solely for economic reasons and rarely get accepted on smallholding farms, e.g., due to the economic implications of losing a cash crop. A system or practice lacking popularity does not mean that it is not sustainable. For example, having a green manuring crop over a cash crop can be argued as ecologically sustainable practice in the majority of systems, yet is not widely adopted mainly for economic reasons. Thus, it becomes the responsibility of all stakeholders to figure out the best possible ways where ecologically sustainable practices can be introduced in a system without losing the sight of the economic sustainability of systems.

Transferability of findings at landscape level

LTEs can answer many questions at field scale or even at regional level, but some issues related to landscape level (e.g., soil erosion and biodiversity) can be better answered through other means. Apart from the limited use of LTEs in answering these questions, transferability of findings from the LTEs to landscape level is also a challenging task. Recently, attempts have been made to address this challenge by coordinating a large number of LTEs in networks at national (Romanenkov et al., 2020; Siebielec et al., 2020; Grosse et al., 2020), continental (Corraddo et al., 2020), and even global level (Macdonald, 2020).

Nevertheless, achieving a yield similar to the LTE level at farmers' field is often under question. Yield variation on farmers' fields is much larger, and researchers always relate this to variation in farm management. On the other hand, farmers tend to argue against this claim of farm management and rather associate it with the difference in resources (human and inputs) per unit area available with researchers compared with that of farmers. One way to address this problem is by following a coordinated approach of having multiple on-farm trials in parallel to LTE and having regular discussions among researchers and farmers, on the findings as well as implementation decisions. This could even be institutionalized by installing advisory committees comprising farmers and scientists. For example, two meetings per annum of DOK advisory board create opportunity for practitioners' help to bring new perspectives on different dimensions of scientific concepts and set a ground for a common language. Similarly, in SysCom trials, this has been achieved through the international scientific board, national-level scientific boards, and regional farmer advisory committees.

Financial resources

Two most crucial assets for survival of any LTE are the availability of operational funds and guaranty of land tenure on which LTE is set up. The former is more critical than latter, as the latter can mostly be solved through the former one. Virtually, all the ongoing LTEs have faced financial challenges at a point of time. Depending upon the institutional settings, different funding structures are adopted to sustain these long-term sites that demand consistent funding. For instance, applying a top to bottom approach, a change initiated at the political level ensured the support and initiation of the systems comparison trial (DOK) in Switzerland, whereas for Rothamsted trials, initial funds were provided privately by Lawes through the establishment of Lawes Agricultural Trust, and it took almost half a century to receive any external funding. In some of the developing countries, many trials are financed by the government with fixed annual budgets, which become barely enough to maintain the trials over time as the operational costs of trial increase. Lack of resources could also mean limited availability of expertise for interpretation and dissemination of results, and many times findings do not get disseminated beyond the annual reports. There are also examples of trials where field activity during financial crunch was limited to just barely maintaining the trial until the funding became available again. With increasing age of LTE, costs of maintaining data archives and sample storage continue to increase; at the same time, repetitiveness of results over the years and slow outputs, e.g., associated to changes in soil properties, increase the risk of lost interest by donner bodies. Therefore, custodians of an LTE not only need to work on the dissemination of findings, but they also need to actively engage with relevant stakeholders for maintaining their interest in scientific value and strategic worth of the LTE. A successful example of banking on excellent network and community engagement for this purpose comes from Russel ranch, where private donors were offered with the opportunity to adopt individual trial plots (Tautges and Scow, 2020). Given the complexity of administrative structures and governance policies as well as the length of time for which financial support is needed, it is always better to have more than one channels with funding potential.

Traps, pitfalls, and precautions

It is relatively comfortable to get convinced and impressed by the value of an LTE. Still, before initiating a new LTE and while interpreting the results, several factors need to be considered and thoroughly evaluated. While any experiment needs to be well planned, LTEs being a long-term commitment demand extra caution and more elaborate planning. In this section, we summarize some key consideration before setting up an LTE:

While there could be several pondering factors working simultaneously to embark on an LTE as soon as possible, the time invested before setting up the trial will certainly pay back later on. Site selection is arguably one of the most crucial aspects, as the location of an LTE could not be changed later on without losses. For logistical reasons, approachability and accessibility of the field site should be given a high priority. As the trial develops, researchers and students need to visit the trial frequently for observations. Also, sufficient infrastructure for storage and processing of trial samples need to be part of the plans, particularly because the volumes continue to increase over the years. Consideration of security perspective, e.g., from wild animals is also an aspect relevant for site selection.

Upon preliminary site selection, determination of homogeneity of the field site is of utmost importance. If possible, more than one sites could be preselected and tested for homogeneity. For instance, the location of DOK trial had to be changed due to strong soil heterogeneity across the experimental site. Fortunately, the issue was realized and resolved within a year of trial setup (Besson et al., 1978). While testing the homogeneity, it is advisable that the soil tests are not limited to plow layer only; rather, elaborate root-zone profiling for physical and chemical properties, depending upon the nature of the trial/crops, will prove useful. The efforts invested there could offer two main advantages: (1) establishing a solid baseline is very important, for which there will be no other possibility later on, and (2) this analysis could be a lifeline for potential extrapolative and explorative studies such as computer modeling.

In practice, the variation in plot size is far-flung among different LTEs across the globe. Some of the factors affecting plot size include type of crop, objective of the trial, and availability of resources (including land and funds). Having a plot size as large as possible could offer certain advantages such as the possibility of split treatments, in case needed at a later stage, and conduct of certain studies, e.g., pollinators, insect pests, and aboveground biodiversity. However, maintaining a trial of substantial size may strain resources over longer term, and therefore, the trade-off between advantages and liabilities needs to be appropriately weighed. One critical point is to have sufficient buffer space among the treatment plots, which would help, e.g., to minimize the drift effect from input application, minimize movement of pests, and provide maneuvering space for mechanical operation. In addition, this space could also be useful as pathways for field visits, since LTEs tend to gain some sort of monumental status over time and attract frequent visits from academics, farmers, and other stakeholders. LTEs can serve as a valuable resource for capacity development and wherever possible logistical as well as institutional arrangements to utilize this potential should already be put in place at the beginning of the trial (Carkner et al., 2020).

For statistical reasons, randomization and replication of treatments are highly relevant; however, it should be weighed against the practical application. For instance, the true randomization in the Chertsey Establishment Trial was not fully implemented, due to the limitation of movement of commercial-scale machinery, to realise the effect of irrigation treatments (Brooker et al., 2020). To ensure the reproducibility of findings, the replication of treatments is essential. For example, the findings of the model farm project of Punjab Agricultural University offer a way-out to local farming from prevalent rice–wheat monoculture (Walia and Kaur, 2020), but additional replications could have enhanced the impact of results even further.

Many of the LTEs are designed around a specific crop-specific and/or production system-specific approach, which might be driven by the interests of a particular funding body. These trials run the potential risk of losing financial support, e.g., if the particular commodity becomes economically uninteresting or the funding body deems that trial has met their trial objectives or looses interest in the trial outcomes over time. Therefore, having a trial integrated into the institutional setup could be of value for sustaining the trial over long term. Multidisciplinary and transdisciplinary approaches could offer excellent opportunities to utilize the potential of the trial fully and to have a broader interest in the trial. Despite the scientific objectives and ambitions, the farmers, being the end users of the trial outcomes, should never get out of focus.

Strategy for stakeholder engagement in long term experiments

To ensure the possibility of extrapolation of results in time and space, while maintaining the relevance for local farming, integration of LTE with a component of participatory on-farm research (POR) is advisable (Fig. I.1). An interdisciplinary

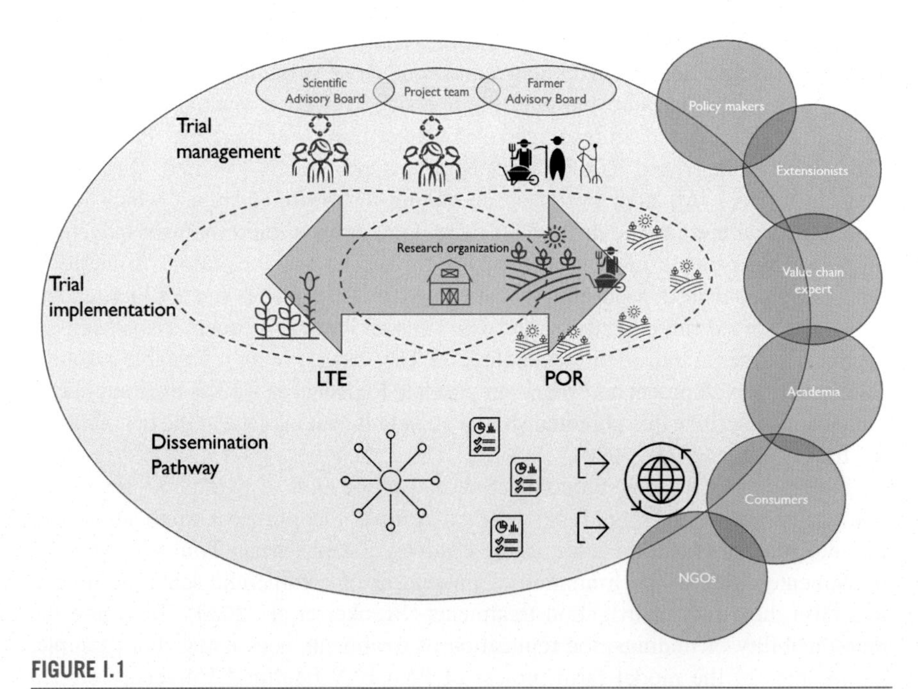

FIGURE I.1

A proposed strategy to manage semistatic and dynamic long-term trails.

and farming systems approach could be adopted to implement POR field trials at a local/regional level. In this approach, the knowledge generated from LTE can be transferred to POR, and standardised best practices and technologies from POR could be implemented in LTE. This enhances the cohesion between on-station and on-farm activities. Such an approach, on the one hand, will facilitate finding practical solutions for real-time farming challenges and, on the other hand, will help maintain the continuity of LTE, as only the practices pretested in POR will be incorporated in LTE. This integration with on-farm research helps in keeping the LTE relevant and exciting to the local farming context and presents an updated and real comparison of different production systems. However, this approach also has its own challenges, e.g., from the LTE perspective, POR findings may only be incorporated at the start of a crop rotation, and not all the results from POR may be directly applicable for LTE. So both LTE and POR trials, in this context, need to be designed carefully considering the aspects that need to be evaluated in on-farm field trials and their way back into LTE. SysCom trials of FiBL (Switzerland) in Kenya and India present good examples of such an integrated approach. The trials were set up in 2007 with an objective of comparing the performance and potential of conventional and organic agricultural production systems in the tropics. The LTEs in each country focus on the most prevalent cropping system of the region and have a strong integration with participatory on-farm research. Thanks to this approach, these LTEs are as popular with local stakeholders as they are deemed valuable by scientific community. A potential outline for management and implementation of an LTE in cooperation with various stakeholders is presented in Fig. I.1. Furthermore, it could be advisable to develop a constitution for trial management defining roles and responsibilities and with clear delineation between decision-making and advisory roles.

References

Ashcroft, P.M., Catt, J.A., Curran, P.J., Munden, J., Webster, R., 1990. The relation between reflected radiation and yield on the Broadbalk winter wheat experiment. Remote Sensing 11 (10), 1821–1836.

Besson, J.-M., Vogtmann, H., Lehmann, V., Augstburger, F., 1978. DOK: Versuchsplan und erste Ergebnisse eines Projekts zum Vergleich von drei verschiedenen Anbaumethoden. Schweizerische landwirtschaftliche Forschung (22), 20–28. Band 17, Heft3/4.

Brooker, T., Poole, N., Chynoweth, R., Horrocks, A., Rolston, P., Pyke, N., 2020. Challenges of maintaining relevance to current agricultural issues in a long-term cropping establishment experiment in Canterbury, New Zealand. In: Bhullar, G.S., Riar, A. (Eds.), Long-term Farming Systems Research: Ensuring Food Security in Changing Scenarios. Academic Press pp (71–88).

Carkner, M., Bamford, K., Martens, J.T., Wilcott, S., Stainsby, A., Stanley, K., Dick, C., Entz, H.M., 2020. Building Capacity from Glenlea, Canada's oldest organic rotation study. In: Bhullar, G.S., Riar, A. (Eds.), Long-term Farming Systems Research: Ensuring Food Security in Changing Scenarios. Academic Press pp (103–122).

Corrado, C., Danilo, C., Daniele, A., Stefano, C., 2020. Long-term experiments on agroecology and organic farming: the Italian LTE network. In: Bhullar, G.S., Riar, A. (Eds.), Long-term Farming Systems Research: Ensuring Food Security in Changing Scenarios. Academic Press pp (183–196).

Forster, D., Andres, C., Verma, R., Zundel, C., Messmer, M.M., Mäder, P., 2013. Yield and economic performance of organic and conventional cotton-based farming systems—results from a field trial in India. PLoS One 8 (12).

Grosse, M., Hoffmann, C., Specka, X., Svoboda, N., 2020. Managing long-term experiment data: a repository for soil and agricultural research. In: Bhullar, G.S., Riar, A. (Eds.), Long-term Farming Systems Research: Ensuring Food Security in Changing Scenarios. Academic Press pp (167–182).

Johnston, A.E., Poulton, P.R., 2018. The importance of long-term experiments in agriculture: their management to ensure continued crop production and soil fertility; the Rothamsted experience. European Journal of Soil Science 69 (1), 113–125.

Krause, H.M., Fliessbach, A., Mayer, J., Mäder, P., 2020. Implementation and management of the DOK long-term system comparison trial. In: Bhullar, G.S., Riar, A. (Eds.), Long-term Farming Systems Research: Ensuring Food Security in Changing Scenarios. Academic Press pp (37–51).

Lawes, J.B., 1882. The future of agricultural field experiments. Agricultural Students Gazette, New Series 1, 33–35.

Macdonald, A.J., 2020. Long-term agricultural research at Rothamsted. In: Bhullar, G.S., Riar, A. (Eds.), Long-term Farming Systems Research: Ensuring Food Security in Changing Scenarios. Academic Press pp (15–36).

Musyoka, M.W., Adamtey, N., Muriuki, A.W., Cadisch, G., 2017. Effect of organic and conventional farming systems on nitrogen use efficiency of potato, maize and vegetables in the Central highlands of Kenya. European Journal of Agronomy 86, 24–36.

Romanenkov, V.A., Shevtsova, L.K., Rukhovich, O.V., Belichenko, M.V., 2020. Geographical network: legacy of the soviet era long-term field experiments in Russian agriculture. In: Bhullar, G.S., Riar, A. (Eds.), Long-term Farming Systems Research: Ensuring Food Security in Changing Scenarios. Academic Press pp (147–165).

Siebielec, G., Matyka, M., Łopatka, A., Kaczyński, R., Kuś, J., Oleszek, W., 2020. Testing long term impact of agriculture on soil and environment in Poland. In: Bhullar, G.S., Riar, A. (Eds.), Long-term Farming Systems Research: Ensuring Food Security in Changing Scenarios. Academic Press pp (123–146).

Tautges, N., Scow, K., 2020. Pursuing agroecosystem resilience in a long-term mediterranean agricultural experiment. In: Bhullar, G.S., Riar, A. (Eds.), Long-term Farming Systems Research: Ensuring Food Security in Changing Scenarios. Academic Press pp (53–69).

Walia, S.S., Kaur, T., 2020. Integration of efficient farm enterprises for livelihood security of small farmers. In: Bhullar, G.S., Riar, A. (Eds.), Long-term Farming Systems Research: Ensuring Food Security in Changing Scenarios. Academic Press pp (89–100).

SECTION

Lessons for agricultural development

CHAPTER

Long-term agricultural research at Rothamsted

1

Andy J. Macdonald[1], Paul R. Poulton[1], Margaret J. Glendining[2], David S. Powlson[1]

[1]*Rothamsted Research, Sustainable Agriculture Sciences Department, Harpenden, Herts, United Kingdom;* [2]*Rothamsted Research, Computational and Analytical Sciences Department, Harpenden, Herts, United Kingdom*

Introduction

Agricultural research began at Rothamsted in the mid-19th century when John Bennet Lawes, the then owner of the Rothamsted estate, established a series of what became long-term field experiments. Lawes understood the importance of soil fertility for crop production and patented a method to manufacture superphosphate fertilizer using crushed animal bones and sulfuric acid, thus increasing the solubility of the phosphate present in bones as calcium phosphate. In 1843, he established a factory on the banks of the River Thames at Deptford, east of London, to manufacture and distribute his new fertilizer (Dyke, 1993). In addition, he sought to improve agricultural prosperity through research to develop independent and unbiased advice for farmers on crop nutrient requirements, including nitrogen (N), phosphorus (P) and potassium (K), by starting two large-scale field experiments on turnips and wheat. He recognized the importance of such knowledge for improving crop production and minimizing the UK dependence on cheap imports of cereals from abroad (especially from the United States and Australia) and to ensure UK food security for a growing urban population. As the fertilizer industry developed the use of fertilizers in agriculture grew, with phosphate fertilizer use in the UK exceeding that of N and K until about 1960 (Johnston and Poulton, 2009).

In 1843, Lawes appointed Joseph Henry Gilbert to help manage the experiments he had planned and to establish further field experiments to test the effects of different mineral fertilizers and organic manures on the yield and quality of other important UK crops, a key aim being to identify which plant nutrients were most limiting for crop production and the amounts required. Together, they established several long-term experiments (LTEs) between 1843 and 1856, of which they abandoned only one, in 1878. Some treatments were changed during the first few years and, later, further changes were made to answer specific questions raised by the results. Further modifications have been made to the experiments since Lawes died in 1900, but seven of the original "Classical" experiments continue today (Table 1.1), albeit in modified forms. They are the oldest, continuous agronomic experiments in the world (Johnston, 1994), and Broadbalk is the oldest continuous

Long-Term Farming Systems Research. https://doi.org/10.1016/B978-0-12-818186-7.00002-3
Copyright © 2020 Elsevier Inc. All rights reserved.

Table 1.1 The Rothamsted Classical Field Experiments and the year they were started.

Year	Experiment
1843	Barnfield—continuous root crops[a]
1843	Broadbalk—continuous winter wheat
1852	Hoosfield—continuous spring barley
1854	Garden Clover—continuous red clover
1856	Park Grass—permanent grassland cut for hay
1856	Alternate Wheat and Fallow[b]
1856	Exhaustion Land—long-term arable cropping

[a] *Currently under grass; treatments not applied since 2000.*
[b] *Modified in 2015; continuous wheat with a small amount of N applied since 2015.*

scientific experiment in existence. Here, we focus on three of the most well-known experiments: the Broadbalk Wheat, Hoosfield Continuous Spring Barley, and Park Grass Continuous Hay Experiments. Details of the other experiments are given in Johnston (1994) and Macdonald et al. (2018). Some of the challenges influencing the management of long-term field experiments and the issues influencing decisions on whether to make changes are discussed by Powlson et al. (2014) and Johnston and Powlson (1994).

Experimental design and management

The primary objectives of many of the Classical experiments were to measure the effects on crop yields of inorganic compounds containing N, P, K, sodium (Na), and magnesium (Mg), elements known to occur in considerable amounts in crops and farmyard manure (FYM), but whose separate actions as plant nutrients had not been studied systematically. The materials used were superphosphate (first made at Rothamsted by treating animal bones with sulfuric acid) and the sulfates of K, Na, and Mg (often referred to then as minerals), and ammonium salts and sodium nitrate (as alternative sources of nitrogen). These inorganic fertilizers were tested alone and in various combinations. Nitrogen was often applied at two or more rates (Macdonald et al., 2018). Their effects were compared with those of FYM and rape cake (produced by pressing rape seed) in most of the experiments, many of which were sited on the flinty silty-clay-loam at Rothamsted (Avery and Catt, 1995; Chromic Luvisol, IUSS Working Group, FAO, 2015).

Growing the same crop each year on the same land was a feature of many of the experiments. Considered bad farming in the 19th century, Lawes and Gilbert reasoned that it was a simplification required to better understand individual crop nutrient requirements. The yields of all produce harvested were recorded and samples of crops and soils were kept for chemical analyses. In addition, from 1866, nutrient losses in drainage were measured on Broadbalk. Daniel Hall, in 1903–06, added a few plots to Broadbalk, Park Grass, and Barnfield, mainly to test the effects of withholding P on plots given NKNaMg, which had been omitted from these experiments (Dyke, 1993). Hall also instigated the first regular liming scheme on Park Grass, the only Classical experiment not sited on a neutral or slightly calcareous soil. Most of the arable experiments are on fields that had previously received the traditional heavy dressings of locally dug chalk, a practice not followed on grassland.

By the late 1940s, there was increasing concern that the soils in some plots receiving ammonium sulfate were becoming so acid that yields were adversely affected. On both Broadbalk and Hoosfield soil, pH declined on some treatments to *c*. 5.0 by the early 1950s. Since then, soil pH on these experiments has been maintained at about 7 by differential lime (chalk) applications and a schedule of liming was started to ensure that yields were not constrained by soil acidity. Work to assess the value of the reserves of soil P and K accumulated in some of the experiments also began. Subsequently, in the mid-1960s, the treatments, management, and cropping on the Broadbalk, Hoosfield Barley, and Park Grass experiments were reviewed, and modifications were introduced to ensure that, as far as possible, the experiments remained relevant to farming practice, but without losing their long-term integrity and value as research facilities.

With remarkable prescience, Lawes and Gilbert retained samples of crops and soils taken for chemical analysis once the initial analyses had been completed. Successive generations of scientists at Rothamsted have continued to add to the collection and the resulting Rothamsted Sample Archive now comprises >300,000 samples. This unique resource is of immense value and continues to be used to generate new data stretching back more than 175 years (Macdonald et al., 2015; Perryman et al., 2018).

Meteorological measurements have been made since the 1850s, when Lawes and Gilbert first collected and analyzed rainwater. Temperature records started at Rothamsted in 1873. Annual rainfall averages 704 mm (mean 1971–2000). The average annual mean air temperature recently (1989–2018) (Fig. 1.1) was approximately 1.1°C warmer than the long-term mean of 9.04°C (1878–1988), consistent with the increases in temperature reported in many parts of the world (Hansen and Sato, 2016). Consequently, these data provide invaluable information about the climatic conditions under which the crops have been grown. Since 1992, additional variables, including wet and dry deposition of nutrients (N, S, etc.) and soil chemistry, have been monitored at Rothamsted as part of the work of the Environmental Change Network (ECN; Scott et al., 2015; www.ecn.ac.uk). The Rothamsted Meteorological data are archived together with crop yields, botanical composition, and analytical data for crops and soils in the electronic Rothamsted Archive (e-RA). The e-RA database is continually updated to increase the amount of information included from the LTEs (Perryman et al., 2018) and is freely available on request via the e-RA website (http://www.era.rothamsted.ac.uk).

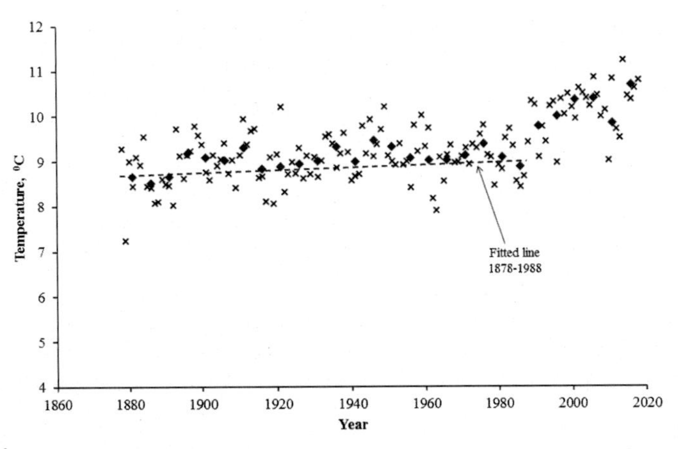

FIGURE 1.1

Average annual temperature (°C) at Rothamsted, 1878–2018. Annual mean (×); 5-year mean (◆).

The Broadbalk Continuous Wheat Experiment

Broadbalk field is thought to have been in arable cropping for many centuries before 1843. A map of 1623 shows it as an arable field, and it is thought that the natural forest vegetation was originally cleared at least 2000 years ago in pre-Roman times. The first experimental crop of winter wheat was sown in autumn of that year and harvested in 1844. Every year since then, wheat has been sown and harvested on all or part of the field. Inorganic fertilizers supplying the elements P, K, Na, and Mg in various combinations and with different amounts of N (initially, 48 (N1), 96 (N2), and 144 (N3) kg N ha^{-1}) were compared with organic manures (FYM and rape cake, later replaced by castor bean meal) and a control treatment that received no fertilizer or manure inputs (Macdonald et al., 2017). These treatments were applied to strips running the entire length of the field, about 320 m long and 6 m wide. For the first few seasons, these treatments were varied a little, but in 1852, a scheme was established that remained largely unaltered until 1968 (Garner and Dyke, 1969). In 1849, a tile drain was laid down the center of each treatment strip 60 cm below the surface, and these led to a main cross drain, which took the water to a soakaway. In 1866, the drains were opened up and drainage water was collected and analyzed. The original drains were still running in the 1990s and were used to make measurements of NO_3–N and P losses. However, they were replaced in autumn 1993 with new perforated plastic pipes, installed 50 cm to one side of the old drain at 75 cm depth, to monitor nutrient losses in drainage from one section only (the current Section 9; see later).

In the early years, the field was plowed using oxen (later by horses) and all the crop from each plot was cut by hand with scythes, bound into sheaves, and carted into the barns to await threshing. Broadbalk is now plowed by a tractor-mounted five-furrow reversible plow and harvested by a small plot combine harvester with a 2 m cut width. Yields of grain and straw are recorded and samples were kept for chemical analysis.

Weeds were initially controlled by hand hoeing. When this became impracticable, owing to a shortage of labor following WW1, five "Sections," (I–V), crossing all the treatment strips at right angles, were made and bare fallowed sequentially (1926–1967). Fallowing was mainly in a 5-year rotation of fallow with four successive crops of wheat, with each phase present each year. The fallow section was cultivated several times to reduce weed populations. FYM and fertilizer were not applied to the fallow phase. Herbicides have been used since 1964 on all of the experiment, except for half of Section V (now Section 8; see later).

Broadbalk has received chalk intermittently since the 1950s to maintain soil pH at a level at which crop yield is not limited. After correction of soil acidity and the introduction of herbicides in the 1950s, a review of the treatments and management led to further modifications being introduced in 1968. The most significant of these were

(i) a change from long-strawed to modern, short-strawed cultivars of wheat with a greater grain yield potential.

(ii) further subdivision of the treatment strips to create 10 new sections (0–9), so that the yield of wheat grown continuously could be compared with that of wheat grown in rotation after a 2-year break. The main purpose of introducing a rotation was to provide a "disease break."

(iii) the replacement of ammonium sulfate and sodium nitrate by ammonium nitrate, initially as "nitro-chalk" (calcium ammonium nitrate), now as "Nitram" (ammonium nitrate), and the introduction of higher N rates (192 (N4) kg N ha^{-1} in 1968; 240 (N5) and 288 (N6) kg N ha^{-1} in 1985). To accommodate these higher N rates, several treatments, thought to be less informative, were stopped.

Sections 0, 1, 8, and 9 continued to grow wheat only, thus maintaining the original cropping practice, with occasional fallows to control weeds on Section 8 which does not receive herbicides (Fig. 1.2). Sections 2, 4, 7 and Sections 3, 5, 6 went into two different 3-course rotations in 1968. Section 6 reverted to continuous wheat in 1978 and the other five sections went into a 5-course rotation; initially fallow, potatoes, wheat, wheat, wheat and from 1997 to 2017, oats (without FYM or N), forage maize, wheat, wheat, wheat. The use of fungicides and insecticides was adopted as part of the management on most sections in the 1970s, as they came into general use in commercial agriculture. Pesticides continue to be applied when necessary, except on Section 6 which does not receive spring or summer fungicides. On Section 0, the straw on each plot has been chopped after harvest and incorporated into the soil since autumn 1986; on all other sections, straw is baled and removed. Plans for the different stages of this and other experiments are available from the e-RA website (http://www.era.rothamsted.ac.uk/Broadbalk).

The experiment is reviewed regularly by an interdisciplinary group, and changes are made when there is a strong scientific case for doing so. An important change, made for the 2000 season, was to withhold P fertilizer from selected plots. This will allow bicarbonate extractable P (Olsen et al., 1954) to decline to more

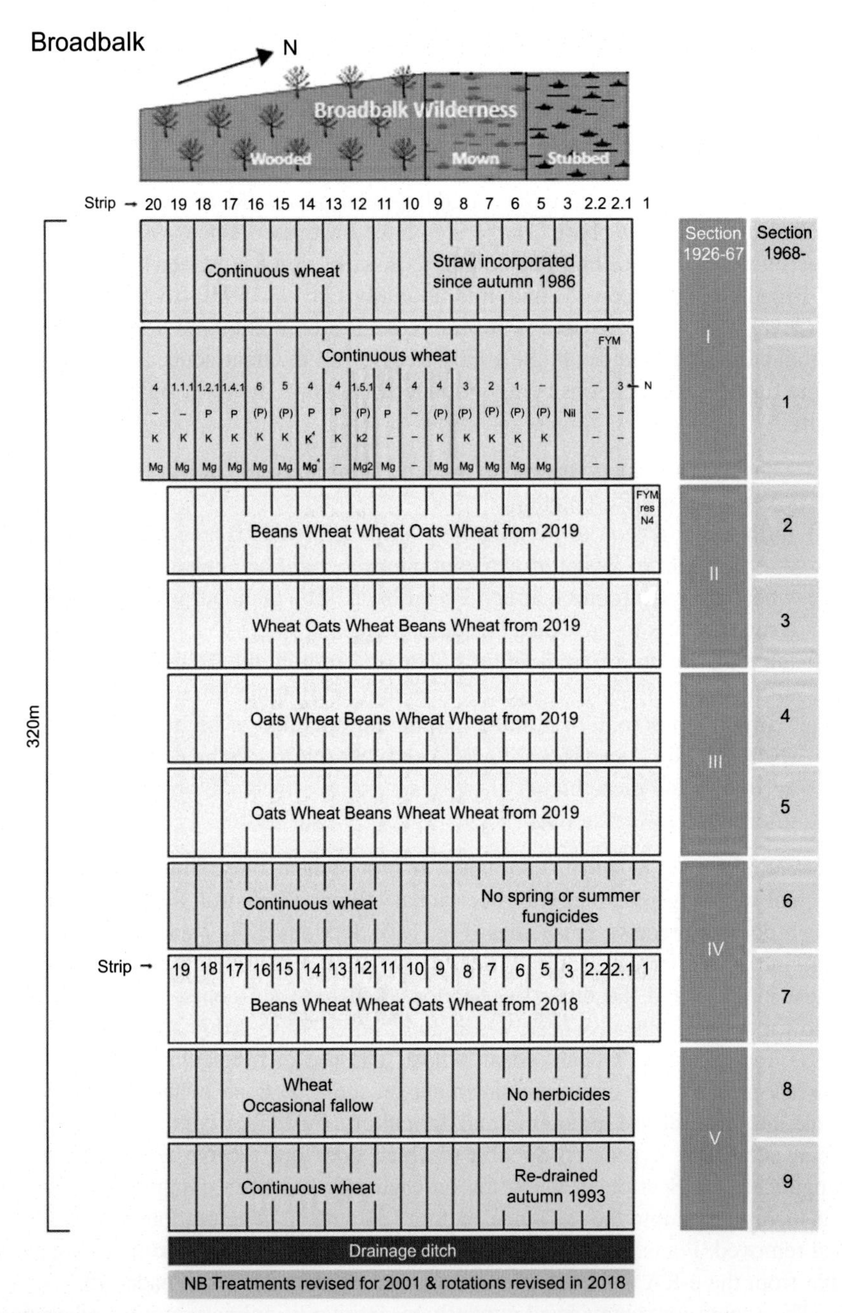

FIGURE 1.2

Plan of the Broadbalk Wheat Experiment.

agronomically realistic levels, especially in low yielding plots that receive small N applications and thus have limited crop P requirements. The unusually high P levels reached in such plots led to large P losses in drainage water (Heckrath et al., 1995). Also, in 2000, treatments on four strips were changed such that a test of split N applications could be included and applications of sulfur-containing fertilizers on strip 14 were stopped to create a "with" and "without" S comparison.

The wheat variety grown is reviewed every 5 years and the crop rotation is reviewed regularly. Consequently, in autumn 2017, winter beans replaced maize in the rotation, and a new rotation of beans (without FYM or N), wheat, wheat, oats, wheat began (Macdonald et al., 2018) (Fig. 1.2); in autumn 2018, *cv.* Zyatt replaced *cv.* Crusoe as the wheat variety grown. Beans were included in the rotation to test their residual nutrient (especially N) value on a subsequent wheat crop. Winter oats, now given N as a single dose at half the usual rates applied to wheat on Broadbalk, were kept as a break crop to help control soil-borne pests and diseases, especially take-all (*Gaeumannomyces graminis* var. *tritici*). The inclusion of two first wheats in the new rotation is designed to enhance the overall productivity of the rotation and examine its longer-term sustainability.

Hoosfield Continuous Spring Barley

Spring barley has been grown continuously on Hoosfield since 1852. It offers interesting contrasts to Broadbalk; being spring-sown, it has only needed to be fallowed four times to control weeds and it tests not only N, P, K, Na, and Mg as well as FYM but also sodium silicate to assess effects on straw strength, which was used in large quantities in the local straw hat factories at that time. The experiment has a factorial design (Warren and Johnston, 1967) with *strips* originally testing four combinations of nutrients: 0 v P v KMgNa v PKMgNa, crossed by four *Series*, originally testing no N or three forms of N, applied (usually) at 48 kg N ha^{-1} (Series 0, no N; Series A, ammonium sulfate; Series AA, sodium nitrate; Series C, rape cake, later castor meal). Additional plots test: unmanured; ashes, 1852–1932; residues of FYM applied 1852–71; FYM since 1852.

Short-strawed cultivars have been grown on the whole experiment since 1968, when most of the existing plots were divided and a four-level N test started (0, 48, 94, 144 kg N ha^{-1}), replacing the test of different forms of N. Ammonium sulfate, sodium nitrate, caster meal, and no N were replaced by ammonium nitrate, initially as "nitro-chalk" (calcium ammonium nitrate), now as "Nitram" (ammonium nitrate). Growing barley in rotation with potatoes and beans was tested on parts of Series AA and C. The effects of the 2-year break on the yield of barley were small, and barley has been grown each year on the whole experiment since 1979. Lime has been applied since 1955 to ensure that soil pH does not limit yield. The straw is baled and removed each year. The cultivation and harvesting methods are largely the same as for Broadbalk, except that in the early years of the experiment, it was cultivated in both autumn and spring, before the crop was sown. Weeds were initially controlled by hand hoeing. Since 1944, weeds have been controlled by herbicides, additional cultivations in the autumn and spring, and hand pulling (Warren and Johnston, 1967). Foliar diseases have been controlled by fungicides since 1978.

More recently, two new plots, started in 2001, test P2KMg and FYM. Strip 5 tested various other combinations of N, P, K, and Mg. In 2003, several major changes were made to the experiment. The four-level N test on the "main" plots continued, but P and Mg are being withheld on some plots (and on parts of Series AA) until levels of plant-available P and Mg in the soil decline to more appropriate agronomic levels. Series C and Strip 5 are now used to test responses to plant-available P; basal N is applied and plots that had not previously received K now receive K fertilizer to ensure that lack of K is not limiting yield. The silicate test on Series AA has been simplified by stopping the four-level N test and applying basal N.

Park Grass

Park Grass is the oldest experiment on permanent grassland in the world (Silvertown et al., 2006). Started in 1856, its original purpose was to investigate ways of improving hay yields by the application of inorganic fertilizers or organic manures (Macdonald et al., 2018). The experiment was established on *c.* 2.8 ha of parkland that had been in permanent pasture for at least 100 years. The uniformity of the site was assessed in the 5 years before 1856. Treatments include controls (Nil—no fertilizer or manure) and various combinations of P, K, Mg, and Na, with various rates of N applied as either sodium nitrate or ammonium salts. FYM was applied to two plots but was discontinued after 8 years because, when applied annually to the surface in large amounts, it had adverse effects on the sward. FYM, applied every 4 years, was reintroduced on three plots in 1905.

The plots are cut in mid-June and made into hay. For 19 years, the regrowth was grazed by sheep penned on individual plots; but since 1875, a second cut, usually carted green, has been taken. The plots were originally cut by scythe, then by horse-drawn, and then tractor-drawn mowers. Yields were originally estimated by weighing the produce, either of hay (first harvest) or green crop (second harvest), and dry matter determined from the whole plot. Since 1960, yields of dry matter have been estimated from one or two strips (each 1.1 m wide) cut within the plots using a box mower. However, for the first cut, the remainder of the plot is still mown and made into hay, thus continuing earlier management and ensuring return of seed. For the second cut, the yield strip is again cut with a box mower and the remaining area is cut with a larger commercial mower. The experiment is never cultivated or resown.

Park Grass probably never received the large applications of chalk that were often applied to arable fields in this part of England. The soil (0–23 cm) on Park Grass probably had a pH (in water) of about 5.5 when the experiment began. A small amount of chalk was applied to all plots during tests in the 1880s and 1890s, and a regular test of liming was started in 1903 when most plots were divided into two and 4 t ha^{-1} $CaCO_3$ applied every 4 years to one half. However, on those plots receiving the largest amounts of ammonium sulfate, this was not enough to stop the soil becoming progressively more acid, making it difficult to disentangle the effects of N from those of acidity. Consequently, it was decided to extend the pH range on each treatment and, in 1965, most plots were divided into four: subplots "a" and "b" on the previously limed halves and subplots "c" and "d" on the previously unlimed halves. Subplots "a," "b," and "c" now receive different amounts of chalk, when necessary, to achieve and/or maintain soil (0–23 cm) at pH 7, 6, and 5, respectively. Subplot "d" receives no lime and its pH reflects inputs from the various treatments and the atmosphere. Soils on the unlimed subplots of the Nil treatments are now at *c.* pH 5.0, while soils receiving 96 kg N ha^{-1} as ammonium sulfate or sodium nitrate are at pH 3.4 and 5.9, respectively. The treatments have thus led to a mosaic of plots having varying soil pH and nutrient status, providing a valuable resource for soil research.

In 1990, plots 9 and 14, which received PKNaMg and N as either ammonium sulfate or sodium nitrate, respectively, were divided so that the effects of withholding N from one half of all the subplots could be assessed. Similarly, plot 13, which received FYM and fishmeal (since 2003 poultry manure), was divided, and, since 1997, FYM and fishmeal have been withheld from one half. In 1996, plot 2, a long-term Nil treatment, was divided and K has been applied to one half each year to give a "K-only" treatment. In 2013, plot 7 was divided into two to test the effects of withholding P on herbage production and botanical diversity. The effects in the three subsequent years were negligible. Consequently, in 2016, the application to all plots receiving P was decreased from 35 to 17 kg P ha^{-1} to more closely match P offtakes. Since 2013, plot 15 has received sodium nitrate at 144 kg N ha^{-1}, in addition to PKNaMg, to provide a comparison with plot 11, which receives the same rate of N as ammonium sulfate.

Results

The early results from the experiments were of immediate importance to farmers, showing which nutrients had the largest effects on the yield of different crops. They showed how N, P, and K accumulated or diminished in soil depending on fertilizer or manure applications, offtakes in crops, and losses in drainage water. The practical value of later results to farmers diminished as the processes of depletion and enrichment of nutrients continued and the processes became well-known. However, since the 1960s, the continuing value of the experiments has become increasingly apparent following changes designed to ensure the experiments remain relevant to current agricultural practice.

Broadbalk

Yields

In his first Rothamsted paper, published over 170 years ago, Lawes described the Broadbalk soil as "a heavy loam resting upon chalk, capable of producing good wheat when well manured" (Lawes, 1847). Similar land nearby, farmed in rotation,

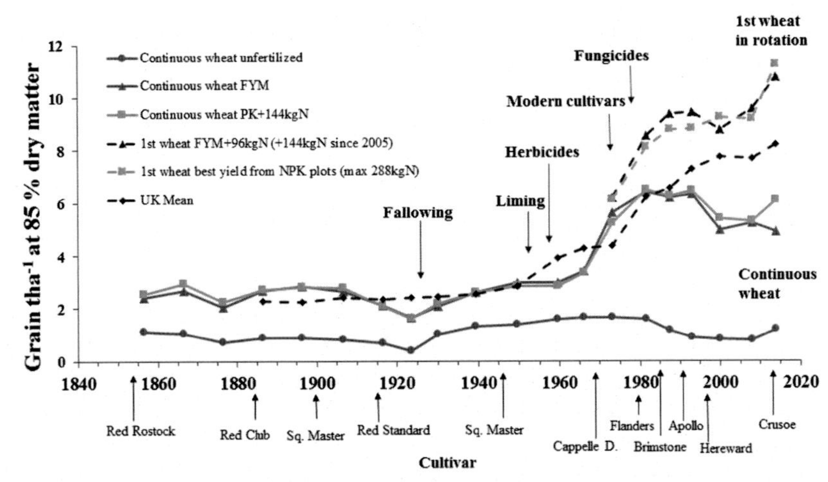

FIGURE 1.3

Mean grain yields of winter wheat (1852–2016) on selected plots of Broadbalk (excluding spring wheat in 2015) and average UK wheat yields (1885–2016), including winter- and spring-sown varieties.

typically yielded *c*.1.2 t ha^{-1}. Fig. 1.3 shows mean yields from selected treatments since the 1850s and also, for comparison, the UK mean wheat yield. The changes reflect the effects of improved cultivars, cultivations, control of pests, diseases, and weeds that have been introduced on Broadbalk, especially since the 1960s. The initial use of continuous cropping (monoculture) clearly demonstrated the value of fertilizer N for wheat production, an effect often masked by the beneficial effect of legumes in earlier rotational experiments. While some of the treatments on Broadbalk underwent changes in the early stages of the experiment, by the 1850s, most of the treatments were established and it provided a convincing visual demonstration of the benefits of inorganic fertilizers, especially N, on the yield of winter wheat. Yields were always greatest on plots treated with N, P, and K together, demonstrating the importance of adequate P for maximum yield. Mean yields of wheat given PKNaMg+144 kg N ha^{-1} were similar to those of wheat given FYM (35 t ha^{-1}, made from cattle manure and containing approximately 220 kg N ha^{-1}), about twice that produced on the unfertilized control (Fig. 1.3).

These early results clearly demonstrated that the supply of N from soil plus deposition from the atmosphere was insufficient to achieve maximum yield for winter wheat, contrary to the views of others at the time. Until *c*.1939, the best wheat yields on Broadbalk were similar to the average yields of wheat grown on UK farms. In the 1950s, with the introduction of higher yielding cultivars and increased use of fertilizers, average wheat yields on UK farms (DEFRA, 2017) exceeded those on Broadbalk until changes to the latter were made in the 1960s (Fig. 1.3). After the change from *cv.* Squarehead's Master to the shorter-strawed cultivar *cv.* Cappelle Desprez in 1968, mean yields of grain on the PKNaMg+144 kg N ha^{-1} and FYM treatments doubled to about 5.4 t ha^{-1}. Since 1968, we have been able to compare the yields of wheat grown continuously and as the first wheat after a 2-year break (Dyke et al., 1983). In the 10 years in which *cv.* Cappelle Desprez was grown, foliar fungicides were not applied and foliar diseases, particularly powdery mildew, were common and most severe on plots given most nitrogen. Since 1979, summer fungicides have been used, when necessary (except on Section 6), and this has allowed us to exploit the greater grain yield potential of modern cultivars (Anon, 1991). The increased responses to N fertilizer in 1979–84 suggested that yields might be greater if larger rates of N were applied, and since 1985, rates of 240 and 288 kg N ha^{-1} have been tested. Yields of *cv.* Brimstone and *cv.* Apollo (1985–1995) in rotation were on average slightly greater, where crops received FYM plus additional N (Fig. 1.3) compared with the best yields from crops given NPK, but with subsequent varieties, *cv.* Hereward and *cv.* Crusoe, there has been no additional benefit of FYM. However, average yields of spring sown forage maize (whole crop) grown in rotation from 2008 to 2012 were greater on plots receiving FYM compared with those given only NPK (Macdonald et al., 2018), perhaps because of improvements in soil physical structure enhancing root development and water/nutrient uptake. These factors are likely to be more significant for a spring-sown crop, with a growing season far shorter than that of autumn-sown wheat (Macdonald et al., 2017). Over time, the annual application of FYM increased the soil organic matter (SOM) content on Broadbalk (Poulton et al., 2018). The plots receiving FYM since 1843 now contain almost three times more organic C in the plow layer (0–23 cm) compared with that in the unfertilized control (Macdonald et al., 2015).

Yields of wheat grown after a 2-year break can be over 2 t ha^{-1} more than yields of continuous wheat, almost certainly because the effects of soil-borne pests and diseases, particularly take-all (*G. graminis* var. *tritici*), are minimized (see later). With *cv.* Crusoe, the largest yields exceeded 13 t ha^{-1} for winter wheat in rotation and were, on average, greater than those with the previous variety (*cv.* Hereward), especially at the larger N rates. While the use of fertilizers, pesticides, and herbicides has enhanced yields on Broadbalk since the experiment began (Fig. 1.3), much of the increase since the late 1960s has been because of the development of wheat varieties with increasing grain yield potential (Austin et al., 1993), especially when grown as a first wheat in rotation. Consequently, the yield increases observed on Broadbalk since 1968 also reflect, but exceed, the general trend seen for wheat in UK agriculture (Fig. 1.3; DEFRA, 2017).

Withholding P fertilizer since 2000 has, so far, had no detrimental effect on yields as plant-available P in the soil still exceeds crop requirements (>Index 3; DEFRA, 2010), but withholding S reduced the average grain yields of first and continuous wheats by 0.6 and 0.2 t ha^{-1}, respectively. This reflects the need for S fertilizer by many UK crops since industrial SO_2 emissions, and thus S deposition, have decreased greatly (Zhao et al., 2003). Surprisingly, compared to single applications of N, applying the same amount of N as three split dressings together with the other agronomic practices used in this experiment did not increase grain yield on this soil type (Macdonald et al., 2018).

Weed, pests, and diseases

Herbicides have been used on all plots since 1964 except on one section (Section 8), reserved for the study of weed population dynamics. No other form of weed control is used on this section except for occasional fallowing and additional cultivations. A comparison of cleaned grain yields from Section 8 with equivalent yields on Section 9, which receives the same fertilizer treatments together with herbicide, shows the importance of effective weed control on maintaining crop yields. In 2014, following 6 years of continuous wheat, yields on plots with adequate available P plus N and K on Section 8 (no herbicide) were on average 81% less (range 59%–91%) than those on Section 9 (with herbicide). There are clear differences in the weed species present on the different plots, largely resulting from differences in amounts of added N. Plot 3, which has never received any fertilizers, is the most diverse (with up to 19 species recorded each year), and species richness declines as the rate of N fertilizer increases (Storkey and Neve, 2018); as few as seven species have been recorded in a given year on Plot 16, which receives the most N (288 kg N ha^{-1}).

Before insecticidal seed dressings were used on Broadbalk, wheat bulb fly (*Delia coarctata*) often caused severe damage to wheat after fallow because bulb fly eggs are laid during the summer on bare soil, and damage is caused by larvae burrowing into the young wheat shoots in the early spring. Plants on soils deficient in K usually suffered most because they had fewer tillers, and damage to the primary shoot often killed the whole plant. The damage was minimized by sowing wheat earlier, but this has resulted in occasional problems with gout fly (*Chlorops pumilionis*; Gratwick M., 1992).

Foliar diseases such as yellow rust (*Puccinia striiformis*), brown rust (*Puccinia triticina*), septoria leaf blotch (*Zymoseptoria tritici*), and powdery mildew (*Blumeria graminis*) are common on the no fungicide section of Broadbalk (Section 6) and differ between years depending on the resistance profile of the wheat cultivar being grown and the weather conditions. The winter wheat cultivar grown on Broadbalk from 2013 to 2018, *cv.* Crusoe, has good resistance against yellow rust, powdery mildew, and Septoria, but is susceptible to brown rust.

Both eyespot (*Oculimacula* spp.) and take-all root disease (*G. graminis* var. *tritici*) are common on Broadbalk, as in many wheat crops in the United Kingdom and northwest Europe. Comparisons of yields and of differences in amounts of take-all between continuous wheat on Broadbalk and that in other fields, growing shorter sequences of cereals, led to the development of the hypothesis of "take-all decline" (Gutteridge et al., 2003). This natural form of biocontrol, where take-all disease becomes less severe in continuously grown wheat compared to its severity in shorter sequences of wheat, is thought to be due to the buildup of antagonistic microflora in the soil. Recent work at Rothamsted has shown that when a low take-all inoculum building wheat cultivar was grown in the first year of wheat cropping, there was a substantial reduction in take-all disease and increases in subsequent grain yields of up to 2.4 t ha^{-1} (McMillan et al., 2018).

Nutrient losses in drainage

The drains on Broadbalk were used to quantify losses of plant nutrients from 1866. They showed minimal loss of P, but larger losses of N and Ca. The average amounts of NO_3–N leached from winter wheat (1990–98) given optimum amounts of N fertilizer (144–192 kg ha^{-1}) were about 30 kg N ha^{-1} (Goulding et al., 2000). Even where no N fertilizer had been applied for more than 150 years, about 10 kg ha^{-1} of NO_3–N was lost each year. Most N was lost where the amount of fertilizer N applied exceeded that needed for "optimum" yield (>192 kg ha^{-1}) or where FYM was applied for many years. Even large annual applications of FYM alone cannot supply enough N for maximum yield because only a proportion of the organic N is mineralized to plant-available forms in the year of application. When FYM is combined with additional fertilizer N, it produces large yields (Fig. 1.3), but can result in very large nitrate leaching losses because considerable nitrate is produced at times when the crop growth is small. Consequently, large manure applications can be regarded as environmentally unsustainable (Macdonald et al., 2017). However, even in years when through drainage was less than average, the EU limit for nitrate in drainage water (11.3 mg N L^{-1}) was sometimes exceeded, where little or no fertilizer N, or FYM, had been applied. Measurements of P (mainly dissolved reactive P) in drainage showed that the critical level, above which the P concentration in the drainage water increased rapidly, was *c.* 60 mg kg^{-1} Olsen P on this soil type (Heckrath et al., 1995).

Hoosfield

Yields of spring-sown barley responded to the change to short-strawed cultivars in 1968, larger amounts of N, and the use of fungicides since 1978, as observed for winter wheat on Broadbalk. The best yields obtained with FYM or FYM + N have increased in line with the more general improvements in UK barley yields seen since the 1940s (Fig. 1.4; DEFRA, 2017). However, in contrast with Broadbalk, weeds were less problematic because the late autumn plowing and spring cultivations on Hoosfield controlled them effectively.

Similarly, yields were less affected by take-all because spring barley is less susceptible than winter wheat. Other results clearly demonstrated the importance of adequate plant-available P and K in soil to maximize N uptake and crop yield. The residual benefits of FYM were also shown when on one plot FYM applications ceased after 20 years. Residues of P and K had accumulated in the soil over this period and were still available for plant uptake 100 years later. Until the 1980s, PK with appropriate amounts of N gave barley yields as large as those with FYM. More recently, with newer spring barley varieties, yields have increased on the long-term FYM soil such that, on average, they are not now matched by fertilizers alone (a maximum of 144 kg N ha^{-1}; Fig. 1.4); the difference is often >2t ha^{-1}.

The difference in yield on these soils, with very different levels of SOM in the top 23 cm (1.0% and 3.8% organic C in NPK and FYM plots, respectively), is probably

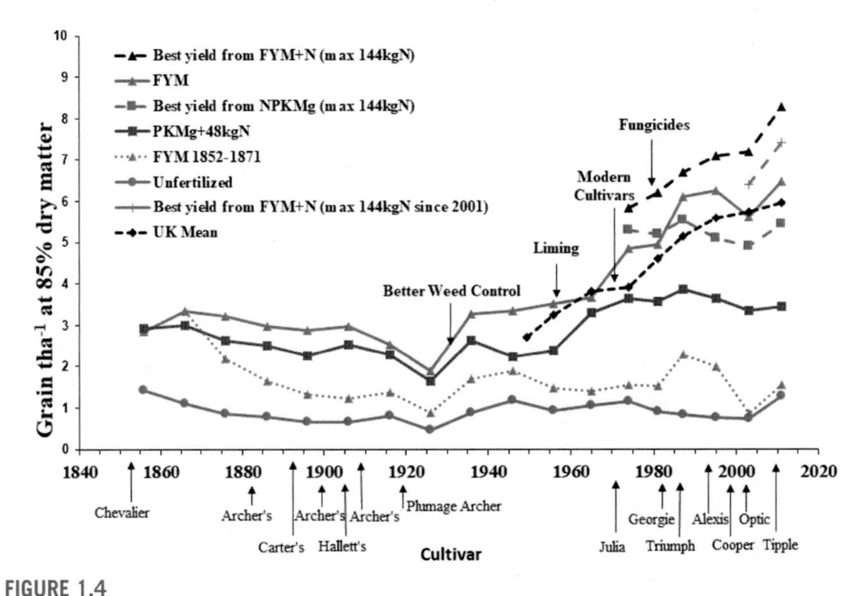

FIGURE 1.4

Mean grain yields of spring barley (1852–2015) on selected plots of Hoosfield and average UK barley yields (1948–2015), including winter- and spring-sown varieties.

because of the improved soil structure and improved water-holding capacity, and to additional N being mineralized and made available to the crop at times in the growing season, and in parts of the soil profile, not mimicked by fertilizer N applied in spring. This contrasts with the effects observed for winter wheat on Broadbalk, where little additional benefit of FYM is observed compared with mineral fertilizers only. However, it is in accordance with the findings for root crops on Barnfield and forage maize on Broadbalk, which showed a clear benefit of FYM on the yields of these spring-sown crops, and also for spring-sown crops in other European LTEs (Hijbeek et al., 2017). To examine this phenomenon in more detail, a new FYM treatment was started on Hoosfield in 2001 to see how quickly yields can be increased and how long it takes for yields comparable with those on the long-continued FYM treatment to be achieved. Yields on the new FYM treatment are now about 2 t ha^{-1} larger than on the NPK plots but are still about 1.0 t ha^{-1} less than those on the long-continued FYM plots (Fig. 1.4). This implies that much of the difference in yield is because of the mineralization of extra N, but there may be further benefits as soil structure gradually improves. However, much of the N mineralized from the extra SOM on the FYM-treated soils is released at a time when it *cannot* be used by the crop and much will be lost by leaching as nitrate (Powlson et al., 1989), as has been found for winter wheat on Broadbalk.

The use of sodium silicate on some plots, both as a fresh application and as a residue, continued to give substantial yield increases in the period 2008–15 on plots lacking P or K but had little effect on plots receiving these nutrients. The mechanism for this is still not fully understood but may be a result of the exchange of silicate for phosphate ions.

Park Grass

Although initial yields recorded on Park Grass were of scientific interest, in practice, most permanent pasture was grazed and yields were of relatively little interest to livestock farmers compared with the effect of grazing on botanical composition. Nevertheless, an analysis of the yields in the 1990s (Anon, 2006) indicated that herbage production had changed little since the experiment began, and all treatments were sustainable. More recent data indicate that on some plots yields have declined (Macdonald et al., 2018).

Effects of fertilizers and lime on herbage yields

The yields of total dry matter (both harvests) for 2012–16 (Fig. 1.5) showed that herbage yields were largest on the limed subplots given PKNaMg and 144 kg N ha^{-1} (N3). Yields with 96 kg N ha^{-1} as either ammonium (N2) or nitrate (N*2) plus PKNaMg are similar; where P or K are not applied, yields are less (Macdonald et al., 2018). Similarly, yields on plots given N only are no better than on the Nil plots because lack of P and K limits yield. Where PKNaMg was applied with no N fertilizer (not shown), yields were as good as those on plots receiving PKNaMg plus 96 kg N ha^{-1} because of the large proportion of legumes in the sward. Where no lime is applied, legumes are less common and yields are smaller. For all treatments, yields on unlimed subplots are less than those on soils maintained at pH 6 or above. Nevertheless, even on the very acid soils (pH 3.4–3.7) dominated by one or two grass species, mean yields can still be as large as 6–7 t ha^{-1}. A comparison of mean yields for 2012–16 with those for 2000–04 (Anon, 2006) indicates that while most yields were similar, those from

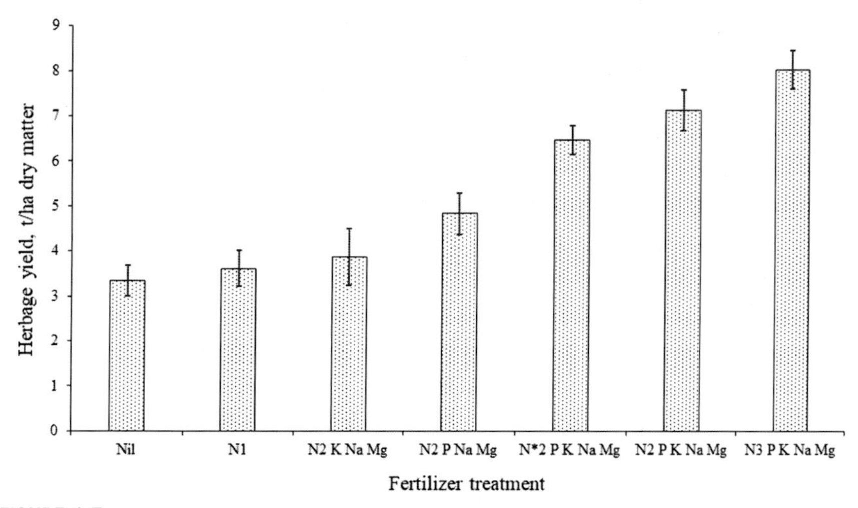

FIGURE 1.5

Mean herbage yields (total of two cuts; ±SE, n = 5) on selected subplots (limed to pH 7) on Park Grass (2012–16).

plots receiving the largest rates of ammonium fertilizers (96–144 kg N ha^{-1}) had declined. However, at various times in the past, there have been periods when yields have declined or increased for some years, for which we have no clear explanation.

Effects of fertilizers and lime on botanical composition

Within 2–3 years of the start of the experiment, the fertilizer treatments had dramatic effects on the species composition of the plots and botanical surveys were started (Johnston, 1994; Crawley et al., 2005). Applications of P and K increased the proportion of legumes in the sward and ammonium fertilizer, with or without P and K, increased the proportion of grasses, and eliminated legumes. The continuing effects on species diversity and on soil function of the original treatments, together with later tests of liming and interactions with atmospheric inputs and climate change, have meant that Park Grass has become increasingly important to ecologists, environmentalists, and soil scientists (Silvertown et al., 2006). Consequently, vegetation surveys have been carried out on Park Grass on more than 30 occasions since the experiments began, and it is a key site within the UK ECN (Scott et al., 2015). The most recent comprehensive surveys of botanical composition, made just before the first cut, were done annually from 1991 to 2000 and from 2010 to 2012. Without exception, all the original treatments imposed at the start of the experiment resulted in a decline in species number; the fertilizers have acted on the initial plant community by selecting out species that are poorly adapted to a higher soil nutrient status or increasing acidity. Even on the unfertilized control the number of species comprising 1%, or more, of the aboveground biomass has decreased since the start of the experiment (Fig. 1.6; Macdonald et al., 2015), possibly as a consequence of atmospheric inputs and/or changes in the management of the sward. Applying either form of N (ammonium or nitrate) decreased species number further in the absence of chalk, but much more so with ammonium sulfate than with sodium nitrate. Applying P alone and PNaMg has decreased the total number of species a little but no more than any other treatment when soils are maintained at pH 5 and above. P applications had relatively minor effects on species composition compared to the Nil treatment, but giving K with P has increased the amount of dry matter from legumes, especially red and white clover (*Trifolium pratense* and *T. repens*) and meadow vetchling (*Lathyrus pratensis*), thus greatly increasing yield. Raising soil pH, by adding chalk, has had bigger effects on the Nil and ammonium sulfate treatments than on those given sodium nitrate (Fig. 1.6).

Uses of the long-term experiments and sample archive

The Classical experiments, Sample Archive and e-RA database, continue to be used to identify and quantify the long-term impact of environmental change and management practices that are impossible to detect in short-term experiments. In recent decades, many of the uses have been dependent on the use of new technologies and could not have been envisaged when the experiments began (Macdonald et al., 2015). The experiments are unique resources available to researchers from

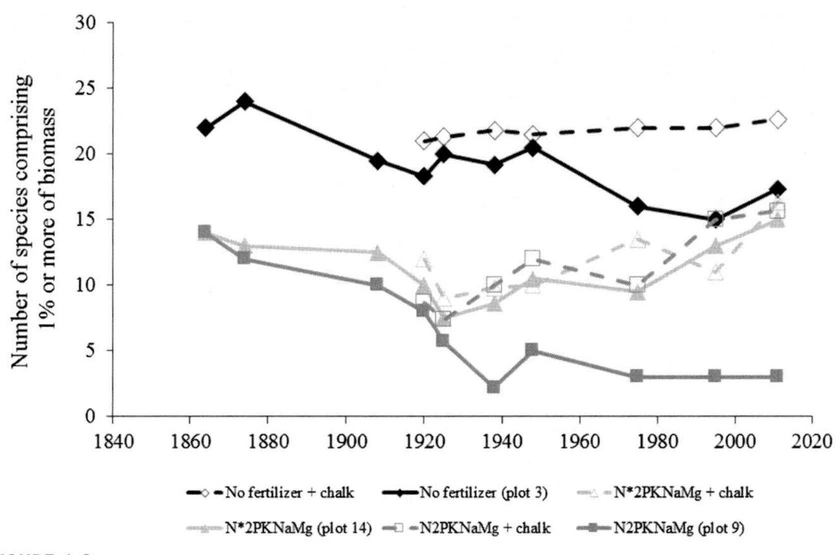

FIGURE 1.6

Changes in the number of species on Park Grass, comprising 1% or more of the aboveground biomass, over time (1864–2011).

all over the world. A few examples of findings from relatively recent studies are given below; many more examples can be found in the e-RA bibliography (http://www.era.rothamsted.ac.uk/papers).

Detailed work on N cycling has been done on Broadbalk, Hoosfield, and Park Grass using ^{15}N-labeled fertilizers (Powlson et al., 1986; Glendining et al., 1997; Jenkinson et al., 2004). Results show that, in our temperate climate, recoveries, by cereals, of fertilizer N greater than 60% can often be achieved and that most of the nitrate present in the soil profile in the autumn, and therefore at risk of loss by leaching, is derived from the mineralization of SOM, not from unused fertilizer N (Macdonald et al., 1989). Exceptions are where excessive amounts of N are applied, in relation to potential crop yield, or where a crop fails because of damage by pests or disease or adverse weather conditions. On Park Grass, labeled N, as either $^{15}NH_4$ or $^{15}NO_3$, was applied in 1980 and 1981. After 18/19 years, 67% of the $^{15}NH_4$–N had been removed in successive grass harvests (mostly in the first year), but a further 17% still remained, in organic forms, in the soil. Less of the $^{15}NO_3$–N was recovered, 60% in the herbage plus 14% in the soil (Jenkinson et al., 2004).

On Park Grass, an increase in legumes, as a percentage of herbage dry matter, has been observed since 2000 on plots where fertilizer N has been withheld since 1989, and also on the Nil treatment. Over the same period, there has also been a marked decrease in atmospheric N deposition largely because of decreases in emissions from industry and road transport (Scott et al., 2015). The results show that grassland species diversity can recover following a decrease in atmospheric pollution and N fertilizer inputs. This provided the first evidence of the impact of anthropogenic stress on biodiversity in an agricultural system followed by recovery after removal of that stress (Storkey et al., 2015).

In other work, Kohler et al. (2012) examined the effects of annual applications of different fertilizers and manures on changes in the intrinsic water-use efficiency (Wi) of the plant communities on Park Grass over a period of nearly 100 years (1915–2009), under conditions of increasing atmospheric CO_2, by measuring the $\delta^{13}C$ enrichment of archived herbage samples from selected plots. Results indicated that carbon isotope discrimination (13Δ) increased significantly ($P < .001$) on the Nil treatment (0.9‰ per 100 ppm CO_2 increase), but this trend differed significantly ($P < .01$) from those observed on the fertilized treatments (PK, N, and NPK). Consequently, the increase of Wi at high N + PK (96 kg N ha^{-1}+PK) was twice that of the control. In addition, the CO_2 responsiveness of 13Δ was related to the grass content of the plant community. This may have been because of the greater CO_2 responsiveness of grasses relative to forbs, indicating that the greater CO_2 response of fertilized swards may be related to the effects of N supply on botanical composition.

Another example of recent work used pyrosequencing of fungal DNA present on archived plant samples from Hoosfield Barley to examine the development of resistance to triazole fungicides in the barley pathogen *Rhynchosporium commune* (leaf blotch fungus). It revealed that for most of the 20th century, the majority of the *R. commune* population on Hoosfield lacked the azole fungicide resistance conferring gene, but in 1985, following the introduction of azole fungicides, levels rapidly increased and subsequently the majority of the *R. commune* population possessed the resistance gene (Hawkins et al., 2014).

Recently, data from 16 LTEs managed by Rothamsted were evaluated to see whether the "4 per 1000 initiative," launched at the Paris Climate Conference in 2015 and aimed at increasing soil organic C content, thus mitigating global warming by C sequestration, is achievable (Poulton et al., 2018). This work showed that while the target of 4‰ per year for 20 years can often be reached by increasing inputs of manure or by major changes in management, such options are not always available to the farmer or may be undesirable for other reasons. In other related work, Broadbalk and Hoosfield have been used to test and develop remote sensing techniques, using drones fitted with multispectral cameras for mapping soil C (Aldana-Jague et al., 2016). ^{14}C-labeling was used to investigate the effect of N fertilizer applications on the activity of methane oxidizing bacteria using soil from plots of contrasting treatment of Park Grass (Stiehl-Braun et al., 2011). In an early example of networking of LTEs, soil C data from seven LTEs globally, including Park Grass, were used to evaluate the performance of nine widely used SOM turnover models (Smith et al., 1997). Similarly, Blake et al. (1999, 2003) investigated climatic and soil effects on K and P balances by comparing data from four different LTEs in different regions of Europe.

Conclusions

The Rothamsted Classical experiments have demonstrated that arable soils in temperate NW Europe can sustain good levels of crop production for generations provided exported nutrients are replaced, soil pH is maintained at around neutral, soil structure is protected, and pests, weeds, and diseases are adequately controlled. They have produced information of value to land users, agronomists, and agroecologists on the management of seminatural systems to maintain and enhance biodiversity. They continue to generate valuable information on the underlying processes upon which our future food security depends (Macdonald et al., 2017) and are a valuable resource for monitoring the impact of human activities on the crop—soil system and the wider environment.

Increasing concerns regarding the wider impacts of climate change on food production systems and vice-versa have highlighted the need for a global network of LTEs to identify the interactions between crops, soils, pests, weeds, and diseases under changing climatic conditions. To this end, a new initiative has been launched at Rothamsted to develop a Global Long-term Experiments Network (GLTEN; https://www.glten.org). As part of this initiative, two new Large-scale Rotation Experiments (LSREs) have been established (Macdonald et al., 2018): one at Broom's Barn (Suffolk, UK) and the other at Rothamsted farm (Herts, UK). It is anticipated that in addition to the existing LTEs at Rothamsted, these new experiments will serve as valuable experimental platforms in the coming years to help design future farming systems with the aim of increasing yields while reducing their impact on the environment.

Acknowledgments

The authors gratefully acknowledge Steve Freeman, Chris Hall, Tony Scott, Andy Gregory, and Sarah Perryman for their support and the many generations of scientific and farm staff (past and present) who have maintained the long-term experiments. The Rothamsted Long-term Experiments are supported by the UK Biotechnology and Biological Sciences Research Council under the National Capabilities program grant (BBS/E/C/000J0300) and by the Lawes Agricultural Trust.

References

Aldana-Jague, E., Heckrath, G., Macdonald, A., Van Wesemael, B., Van Oost, K., 2016. UAS-based soil carbon mapping using VIS-NIR (480—1000 nm) multi-spectral imaging: potential and limitations. Geoderma 275, 55—66.

Anon, 1991. Guide to the Classical Field Experiments. Lawes Agricultural Trust and Rothamsted Experimental Station, Harpenden, Herts, UK, ISBN 0951445626.

Anon, 2006. Guide to the Classical and Other Long-Term Experiments, Datasets and Sample Archive. Lawes Agricultural Trust Co. Ltd. Rothamsted Research, Harpenden, Herts, UK, ISBN 0951445693.

Austin, R.B., Ford, M.A., Morgan, C.L., Yeoman, D., 1993. Old and modern wheat cultivars compared on the Broadbalk wheat experiment. European Journal of Agronomy 2 (2), 141—147.

Avery, B.W., Catt, J.A., 1995. The Soil at Rothamsted. Lawes Agricultural Trust Co. Ltd, Harpenden UK, 44 + map.

Blake, L., Mercik, S., Koerschens, M., Goulding, K.W.T., Stempen, S., Weigel, A., Poulton, P.R., Powlson, D.S., 1999. Potassium content in soil, uptake in plants and the potassium balance in three European long-term field experiments. Plant and Soil 216 (1–2), 1–14.

Blake, L., Johnston, A.E., Poulton, P.R., Goulding, K.W.T., 2003. Changes in soil phosphorus fractions following positive and negative phosphorus balances for long periods. Plant and Soil 254, 245–261.

Crawley, M.J., Johnston, A.E., Silvertown, J., Dodd, M., de Mazancourt, C., Heard, M.S., Henman, D.F., Edwards, G.R., 2005. Determinants of species richness in the Park Grass experiment. The American Naturalist 165, 179–192.

DEFRA, 2010. Fertiliser Manual (RB 209), eighth ed. The Stationery Office, London, UK, pp. 1–250. www.defra.gov.uk.

DEFRA, 2017. United Kingdom Cereal Yields 1885–2017. Farming Statistics, Department for Environment, Food and Rural Affairs. https://data.gov.uk/dataset/.

Dyke, G.V., 1993. John Lawes of Rothamsted. Pioneer of Science, Farming and Industry. Hoos Press, Harpenden, 234pp.

Dyke, G.V., George, B.J., Johnston, A.E., Poulton, P.R., Todd, A.D., 1983. The Broadbalk wheat experiment 1968–78: yields and plant nutrients in crops grown continuously and in rotation. Rothamsted Experimental Station Report for 1982, Part 2 5–44. https://doi.org/10.23637/ERADOC-1-34179.

Garner, H.V., Dyke, G.V., 1969. The Broadbalk yields. Rothamsted Experimental Station Report for 1968, Part 2 26–46. https://doi.org/10.23637/ERADOC-1-34917.

Glendining, M.J., Poulton, P.R., Powlson, D.S., Jenkinson, D.S., 1997. Fate of N-15-labelled fertilizer applied to spring barley grown on soils of contrasting nutrient status. Plant and Soil 195 (1), 83–98.

Goulding, K.W.T., Poulton, P.R., Webster, C.P., Howe, M.T., 2000. Nitrate leaching from the Broadbalk Wheat Experiment, Rothamsted, UK, as influenced by fertilizer and manure inputs and the weather. Soil Use & Management 16 (4), 244–250.

Gratwick, M., 1992. Gout fly. In: Gratwick, M. (Ed.), Crop Pests in the UK. Springer, Dordrecht.

Gutteridge, R.J., Bateman, G.L., Todd, A.D., 2003. Variation in the effects of take-all disease on grain yield and quality of winter cereals in field experiments. Pest Management Science 59, 215–224.

Hansen, J., Sato, M., 2016. Regional climate change and national responsibilities. Environmental Research Letters 11, 034009.

Hawkins, N.J., Cools, H.J., Sierotzki, H., Shaw, M.W., Knogge, W., Kelly, S.L., Kelly, D.E., Fraaije, B.A., 2014. Paralog re-emergence: a novel, historically contingent mechanism in the evolution of antimicrobial resistance. Molecular Biology and Evolution 31, 1793–1802.

Heckrath, G., Brookes, P.C., Poulton, P.R., Goulding, K.W.T., 1995. Phosphorus leaching from soils containing different phosphorus concentrations in the Broadbalk experiment. Journal of Environmental Quality 24 (5), 904–910.

Hijbeek, R., van Ittersum, M.K., ten Berge, H.F.M., Gort, G., Spiegel, H., Whitmore, A.P., 2017. Do organic inputs matter – a meta-analysis of additional yield effects for arable crops in Europe. Plant and Soil 411, 293–303.

IUSS Working Group WRB, 2015. World Reference Base for Soil Resources 2014, Update 2015. International Soil Classification System for Naming Soils and Creating Legends for Soil Maps. World Soil Resources Reports No. 106. FAO, Rome.

Jenkinson, D.S., Poulton, P.R., Johnston, A.E., Powlson, D.S., 2004. Turnover of nitrogen-15-labeled fertilizer in old grassland. Soil Science Society of America Journal 68 (3), 865–875.

Johnston, A.E., Poulton, P.R., 2009. Nitrogen in Agriculture: An Overview and Definitions of Nitrogen Use Efficiency. Proceedings International Fertilizer Society 651, York, UK.

Johnston, A.E., Powlson, D.S., 1994. The setting-up, conduct and applicability of long-term continuing experiments in agricultural research. In: Greenland, D.J., Szabolcs, I. (Eds.), Soil Resilience and Sustainable Land Use. CAB International, Wallingford, Oxon, UK, pp. 395–421.

Johnston, A.E., 1994. The Rothamsted classical experiments: in long-term experiments in agricultural and ecological sciences. In: Proceedings of a Conference to Celebrate the 150th Anniversary of Rothamsted Experimental Station, held at Rothamsted, 14–17 July 1993. CAB International Wallingford, Oxon, UK.

Kohler, I.H., Macdonald, A., Schnyder, H., 2012. Nutrient supply enhanced the increase in intrinsic water-use efficiency of a temperate semi-natural grassland in the last century. Global Change Biology 18 (11), 3367–3376.

Lawes, J.B., 1847. On agricultural chemistry. Journal of the Royal Agricultural Society of England 8, 226–260.

Macdonald, A.J., Poulton, P.R., Clark, I.M., Scott, T., Glendining, M.J., Perryman, S.A.M., Storkey, J., Bell, J., Shield, I., McMillan, V., Hawkins, J., 2018. Guide to the Classicals and other Long-term Experiments, Datasets and Sample Archive. Rothamsted Research, Harpenden, Herts, ISBN 978-1-9996750-0-4, pp. 1–57. www.rothamsted.ac.uk.

Macdonald, A.J., Poulton, P.R., Glendining, M.J., Goulding, K.W.T., Perryman, S.A.M., Powlson, D.S., 2017. Sustainable intensification – lessons from the Rothamsted long-term experiments. Aspects of Applied Biology, Sustainable Intensification 136, 239–244.

Macdonald, A.J., Powlson, D.S., Poulton, P.R., Watts, C.W., Clark, I.M., Storkey, J., Hawkins, N.J., Glendining, M.J., Goulding, K.W.T., McGrath, S.P., 2015. The Rothamsted long-term experiments. Aspects of applied biology. Valuing Long-term Sites and Experiments for Agriculture and Ecology 128, 1–10.

Macdonald, A.J., Powlson, D.S., Poulton, P.R., Jenkinson, D.S., 1989. Unused fertiliser nitrogen in arable soils its contribution to nitrate leaching. Journal of the Science of Food and Agriculture 46, 407–419.

McMillan, V.E., Canning, G., Moughan, J., White, R.P., Gutteridge, R.J., Hammond-Kosack, K.E., 2018. Exploring the resilience of wheat crops grown in short rotations through minimising the build-up of an important soil-borne fungal pathogen. Nature Scientific Reports 8, 9550. https://doi.org/10.1038/s41598-018-25511-8.

Olsen, S.R., Cole, C.V., Watanabe, F.S., Dean, L.A., 1954. Estimation of Available Phosphorus in Soils by Extraction with Sodium Bicarbonate. United States Department of Agriculture Circular No 939.

Perryman, S.A.M., Castells-Brooke, N.I.D., Glendining, M.J., Goulding, K.W.T., Hawkesford, M.J., Macdonald, A.J., Ostler, R.J., Poulton, P.R., Rawlings, C.J., Scott, T., Verrier, P.J., 2018. The electronic Rothamsted Archive (e-RA): a unique online resource for data from the Rothamsted long-term experiments. Nature Scientific Data 5, 180072.

Poulton, P., Johnston, J., Macdonald, A., White, R., Powlson, D., 2018. Major limitations to achieving "4 per 1000" increases in soil organic carbon stock in temperate regions: evidence from long-term experiments at Rothamsted Research, UK. Global Change Biology 24 (6), 2563–2584. https://doi.org/10.1111/gcb.14066.

Powlson, D.S., Macdonald, A.J., Poulton, P.R., 2014. The continued value of long-term field experiments: insights for achieving food security and environmental integrity. In: David Dent) (Ed.), Soil as World Heritage. Springer, Dordrecht, pp. 131–157.

Powlson, D.S., Pruden, G., Johnston, A.E., Jenkinson, D.S., 1986. The nitrogen cycle in the Broadbalk Wheat Experiment: recovery and losses of 15N-labelled fertilizer applied in spring and inputs of nitrogen from the atmosphere. Journal of Agricultural Science 107 (3), 591–609.

Powlson, D.S., Poulton, P.R., Addiscott, T.M., McCann, D.S., 1989. Leaching of nitrate from soil receiving organic or inorganic fertilizers continuously for 135 years. In: Hansen, J.A.A., Henriksen, K. (Eds.), Nitrogen in Organic Wastes Applied to Soils. Academic Press, pp. 334–345.

Scott, T., Macdonald, A., Goulding, K., 2015. The U.K. Environmental change network, Rothamsted. Physical and atmospheric measurements: the first 20 Years. Lawes Agricultural Trust 32. https://www.rothamsted.ac.uk/sites/default/files/PDF-Booklet-2-ECNRothamsted First20yr_revised2015.pdf. isbn:978-0-9564424-1-3.

Silvertown, J., Poulton, P., Johnston, E., Edwards, G., Heard, M., Biss, P.M., 2006. The Park grass experiment 1856–2006: its contribution to ecology. Journal of Ecology 94, 801–814.

Smith, P., Smith, J.U., Powlson, D.S., McGill, W.B., Arah, J.R.M., Chertov, O.G., Coleman, K., Franko, U., Frolking, S., Jenkinson, D.S., Jensen, L.S., Kelly, R.H., Klein-Gunnewiek, H., Komarov, A.S., Li, C., Molina, J.A.E., Mueller, T., Parton, W.J., Thornley, J.H.M., Whitmore, A.P., 1997. A comparison of the performance of nine soil organic matter models using datasets from seven long-term experiments. Geoderma 81 (1–2), 153–225.

Stiehl-Braun, P.A., Powlson, D.S., Poulton, P.R., Nildaus, P.A., 2011. Effects of N fertilizers and liming on the micro-scale distribution of soil methane assimilation in the long-term Park Grass experiment at Rothamsted. Soil Biology and Biochemistry 43 (5), 1034–1041.

Storkey, J., Macdonald, A.J., Poulton, P.R., Scott, T., Köhler, I.H., Schnyder, H., Goulding, K.W.T., Crawley, M.J., 2015. Grassland biodiversity bounces back from long-term nitrogen addition. Nature 528 (7582), 401–404.

Storkey, J., Neve, P., 2018. What good is weed diversity? Weed Research 58 (4), 239–243.

Warren, R.G., Johnston, A.E., 1967. Hoosfield Continuous Barley. Rothamsted Experimental Station Report for 1966, pp. 320–338. https://doi.org/10.23637/ERADOC-1-1966320.

Zhao, F.J., Knights, J.S., Hu, Z.Y., McGrath, S.P., 2003. Stable sulfur isotope ratio indicates long-term changes in sulfur deposition in the Broadbalk experiment since 1845. Journal of Environmental Quality 32 (1), 33–39.

CHAPTER

Implementation and management of the DOK long-term system comparison trial

2

Hans-Martin Krause[1], Andreas Fliessbach[1], Jochen Mayer[2], Paul Mäder[1]

[1]*Research Institute of Organic Agriculture (FiBL), Department of Soil Sciences, Frick, Switzerland;*
[2]*Agroscope, Department Agroecology and Environment, Zurich, Switzerland*

History and scope of the DOK trial

The DOK system comparison trial compares biodynamic (D), bioorganic (O), and conventional (K) agricultural cropping systems since 1978 (Mäder et al., 2002). The origins and the evolution of the DOK system comparison trial mirror contemporary societal developments. The years around the 1970s were shaped by the first resource crises and perceivable signs of severe environmental pollution. Following the guiding principles of the green revolution, the Swiss agricultural sector had shifted its production system toward a highly mechanized and input-intensive agricultural practice in the 50s and 60s. This enabled high yields at minimized labor costs. On the other hand, contemporary publications like "Silent spring" (Carson, 1962) or "The Limits to Growth" (Meadows et al., 1972) were cautioning the effects of industrialized agricultural practices on resource use, biodiversity, and environmental pollution. In "The Limits of Growth," the finiteness nature of natural resources had been expressed for the first time. This questioned society's perception on land use and sustainable food production. The increasing societal concerns on environmental impact of agricultural practices stimulated the development of organic agricultural movements across the globe. Organic agricultural practice negates chemical inputs for plant nutrition and protection, but rather focuses on balancing ecosystem services provided by agricultural land through diverse crop rotations, indirect pest, disease and weed control measures, and nutrient cycling (Seufert and Ramankutty, 2017; Reganold and Wachter, 2016). Yet, during the booming phase of conventional agriculture, the organic agricultural movement was widely perceived as "outdated" and "unfeasible." Especially towing to the issue of yield gaps (Seufert et al., 2012), organic agricultural was challenging the paradigm of the green revolution at a time when increasing market forces demanded instant high productivity at low labor cost. There was a belief that the organic approach, with its emphasis on soil quality, biodiversity, and long-term sustainability, could not cope with these demands and would deplete soils. Nevertheless, a

Long-Term Farming Systems Research. https://doi.org/10.1016/B978-0-12-818186-7.00003-5
Copyright © 2020 Elsevier Inc. All rights reserved.

small alliance of organic pioneers, scientists, and politicians managed to advance their request: A scientific investigation on the feasibility of organic agricultural practices in Switzerland. Initiated by these actors, in 1971, a member of the Swiss national council successfully proposed a motion to the Swiss parliament that ensured federal support for research on organic agriculture. Backed with political support, which included financial commitment to contribute to future projects, the "Research Institute of Organic Agriculture" (FiBL) was founded in 1974. One year later, the Federal Office of Agriculture instructed the "Federal Research Institute of Agrochemistry and Environmental Hygiene" (FAL) and the "Research Institute of Organic Agriculture" with the implementation of a long-term research trial comparing different approaches of organic and conventional agricultural systems.

From 1975 to 1976, a first field trial was established on the Bruderholzhof in Oberwil (Basel Land) south of Basel (Besson et al., 1978). This trial included three farming systems, consisting of biodynamic, bioorganic, and conventional agricultural practices, and a 5-year experimental crop rotation (winter wheat, cabbage, barley, and 2 years of grass clover). Each crop was cultivated in every year to allow annual comparison of system performance. With three systems, two fertilizer intensities, three replicates, and five crops, the field trial exhibited a total of 90 experimental plots. However, soil analyses carried out in 1975 revealed strong soil heterogeneities across the experimental site with respect to soil texture and base saturation (Besson et al., 1978). In 1975 and 1976, significant effects of initial soil geochemical conditions on cereal and cabbage yields were identified and it was decided that a new, more homogenous experimental site was needed to adequately address the scope of the field trial. In addition, a field with less slope was required to minimize potential crossover effects. An alternative location on the Birsmattenhof in Therwil (Basel Land) in the Leimental was found. Soil analyses on the 4 hectare parcel were conducted in 1976 surveying soil texture (clay, silt, sand), soil pH, and cation exchange capacity. In 1976, oat was sown on the whole parcel to get a first indication of the homogeneity of the site based on visual inspection of crop growth. As a result of these observations, the final position of the 1.5 ha requiring long-term trial was defined and a new set of in-depth soil analyses was conducted. The analyses showed that the field site in Therwil was distinctly more homogenous compared with the one in Oberwil (Besson et al., 1986). To deal with still occurring slight soil heterogeneity of soil geochemical parameters, a block design was established. In the later years of the field trial, a common statistical model was developed, which, among others, includes clay content as a covariable, as well as row, line, and row × line effects. Soil nutrient contents, namely C, N, P, and Ca, at the field site in Therwil exceeded soil nutrient status in Oberwil by far. Consequently, a new management plan with a higher share of high nutrient demanding crops and a 7-year crop rotation was developed to account for higher soil fertility at the final field site. In addition, two new unfertilized control treatments were introduced in the experimental setup and only three crops, out of seven in the rotation, entered the management plan for each year. The control treatments either received chemical pesticides, similar to the conventional systems, or biodynamic preparations to compare with the biodynamic and bioorganic systems. The initial aim of the two control treatments was also to monitor decline in crop yield as consequence of soil nutrient depletion.

The whole planning processes of the experiment under lead of the "Federal Research Institute of Agrochemistry and Environmental Hygiene" and FiBL were supported by experts in various disciplines in the field of agronomy, biology, phytopathology, and statistics at ETH Zurich and University Hohenheim. During the establishment of the field trials in Oberwil and Therwil, practitioners from biodynamic, organic, and conventional farming systems were involved in an advisory board to design management practices for the treatments implemented at the DOK field trial. Until today, this group meets every year at least twice to discuss outcomes and management practices ensuring practical relevance of the field trial. In this meeting also scientists present their work and findings to the practitioners. Often new perspectives on scientific investigations evolve during this exchange and the applicability of the research findings undergoes a first reality check. Although for scientists and practitioners these meetings can be challenging, because of its open and constructive colloquial culture, the interdisciplinary exchange proofed mutually valuable and helped to find a common language for communication.

Owing to the system approach, the management of agricultural practices in the DOK trial inherently is a challenging task. Unlike field trials with full factorial experimental designs, the management practices in a system approach by necessity are not fully defined and a range of interpretations on how treatments should be implemented to reflect farming systems needed to be addressed and discussed. At the time of establishment of the long-term trial, no food label with clear guidelines for "organic" food production existed. The foundation of food labels like VSBLO (later Bio-Suisse, bio-suisse.ch) in 1981, and IP-Suisse in 1989 (ipsuisse.ch), with their guidelines for food production facilitated clear definition of management practices. The evolution of the food production guidelines according to the Bio-Suisse label was closely linked to the establishment and implementation of the DOK trial. In this regard, it should be noticed that clear guidelines for biodynamic food production already evolved decades earlier with the foundation of the Demeter Trade association in 1932 (demeter.ch). Later, the work on Swiss production guidelines was also influencing the guidelines for organic farming in Europe and at international level (Schmid, 2007).

Another rather scientific challenge of the system approach implemented in the DOK trial consists in the interpretation of data originating from different farming systems which differ in manifold management factors. Proving a causal relationship between management practices and a measured parameter is even more difficult, if not impossible, compared with field trials with full factorial design. From the point of view of basic research, the system approach trades practical statements on the outcomes of each farming system and a proximity to actually applied agricultural practice against the possibility to gain mechanistic insights on the mode of functionality within each treatment. Consequently, the experimental setup and the modality of implementation of the field trial manifest the tension between fundamental and applied research. Nevertheless, the system approach, which is the outcome of a holistic view, holds other advantages such as a wide coverage of possible management options for each farming systems. This flexibility enables a scientific comparison of such different approaches like organic and conventional agriculture at the first place (Biggs, 1985).

Experimental setup and management practices

Today the DOK field trial is managed by the joint efforts of the Research Institute of Organic Agriculture (FiBL) and the Federal Swiss Institute for Agricultural Research, Agroscope. It is located ~10 km south of Basel next to the village of Therwil. The river Birsig and the surrounding bedrock of the Table Jura shape the geomorphology of the region. Today the majority of the land in the basin is used for rather intensive agriculture with mainly arable and fodder crops. Although the share of pomiculture in the region decreased in the course of the 20th century, orchards are still an integral part of the cultural landscape. The mild climate, with mean annual temperature of 10.5°C and mean annual precipitation of 842 mm, allows an effective vegetation period of 210–215 days per year.

According to the World reference base for soil resources, the prevailing soil type is classified as haplic luvisol with mean clay and sand contents of 15.7% and 11.7% (Fliessbach et al., 2007). Pedogenesis is strongly influenced by alluvial loess deposits, which cover the first 0.9–1.3 m of soil depth. Superficial decalcification leads to vertical transport of clay minerals with rainwater seepage. Subsequent precipitation of clay minerals in deeper soil depths with higher pH forms a characteristic, clay-enriched, B-horizon inducing temporary water stagnation during wet summer months. According to the regional soil classification system, the soil is classified as "schwach pseudovergleyte Parabraunerde."

An inquiry covered land use dates back to the late 1950s. Before the establishment of the long-term trial, the site was used for arable cropping from 1957 until 1973 including temporary grass clover. Cereals and vegetables followed in 1974 and 1975. In 1976, the field site was already under first investigations and oat was cultivated on 4 hectare to visually assess spatial heterogeneity of crop performance. Grass clover was sown in 1977 on the final field site of 1.5 ha to recuperate and prepare soil for future experimental observation. The field trial started in spring 1978 with the cultivation of potato, wheat, and barley.

The experimental setup of the field trial aims at comparing regionally applied organic and conventional farming systems. Eight treatments were implemented varying in fertilizer intensities, fertilizer types, pest, disease, and weed control measures (Table 2.1). The treatments represent different farming systems and include two organic (BIODYN, BIOORG) and two conventional systems (CONFYM, CONMIN) with an unfertilized control (NOFERT). The treatments BIODYN, BIOORG, and CONFYM are maintained at two fertilization intensities. The farming systems implemented in the treatments of the DOK trial mimic agricultural practices of the food production labels "Demeter" (BIODYN), "Bio-Suisse" (BIOORG), and "IP-Suisse" (CONFYM) (Fliessbach et al., 2007). The treatment CONMIN, another integrated

Table 2.1 Fertilizer types, pest, and weed control implemented in the treatments of the DOK System Comparison Trial.

Farming systems		Organic		Conventional		Control
Treatments	1.4 LU	BIODYN 2	BIOORG 2	CONFYM 2	CONMIN 2	
	0.7 LU	BIODYN 1	BIOORG 1	CONFYM 1		
	0.0 LU					NOFERT
Fertilizer Type		Bio-dynamic	Bio-organic	Conventional with farmyard manure	Conventional with solely mineral fertilization	No fertilization
		Aerobic, biodynamic manure compost and slurry	Slightly aerobically rotted manure, slurry and small amounts of potassium magnesia	Stacked manure and slurry mineral N, P, and K fertilization according to federal guidelines	Mineral N, P, and K fertilization according to federal guidelines	
Weed control		mechanical	mechanical	Mechanical and herbicides	Mechanical and herbicides	mechanical
Disease control		indirect	Indirect, copper	Fungicides (thresholds)	Fungicides (thresholds)	Indirect
Pest control		Plant extracts, biocontrol	Plant extracts, biocontrol	Insecticides (thresholds)	Insecticides (thresholds)	biocontrol
Special treatments		Bio-dynamic preparations	none	Plant growth regulators	Plant growth regulators	none

LU, *livestock units fertilization equivalents. 1 LU corresponds to 105 kg N, 15 kg P, 149 kg K, 12 kg Mg, and 37 kg Ca (Richner and Sinaj, 2017).*

Modified from Mäder, P., Fliessbach, A., Dubois, D., Gunst, L., Jossi, W., Widmer, F., Oberson, A., Frossard, E., Oehl, F., Wiemken, A., Gattinger, A., Niggli, U., 2005. The DOK Experiment (Switzerland). ISOFAR Scientific Series, No 1: Long Term Experiment in Organic Farming. Verlag Dr. Köster Berlin, pp. 41–58.

system, was introduced after the first crop rotation and replaced an unfertilized control system that received chemical plant protection only. The implementation of this treatment shifted the scope and focus of the field trial and is discussed in detail in Changes of systems and management section. The eight treatments are replicated four times in three iterations with three crops each. This results in 96 experimental plots of $5 \times 20\ m^2$. The three iterations cover three entry points of the 7-year crop rotation and were conducted to minimize the effects of seasonal variability on treatment performances such as crop yields and to allow more than one observation per crop in a 7-year crop rotation. The experimental plots with eight treatments and three crops are arranged in a block design. The iterations a, b, and c represent three temporally shifted replications of the whole crop rotation (Fig. 2.1).

Mixed farming systems with livestock and arable crops are typical for Switzerland. Thus BIODYN2, BIOORG2, and CONFYM2 receive farmyard manure and slurry corresponding to 1.4 livestock units per hectare, based on the fodder crops in the rotation. The addition of farmyard manure in CONFYM1, BIOORG1, and BIODYN1

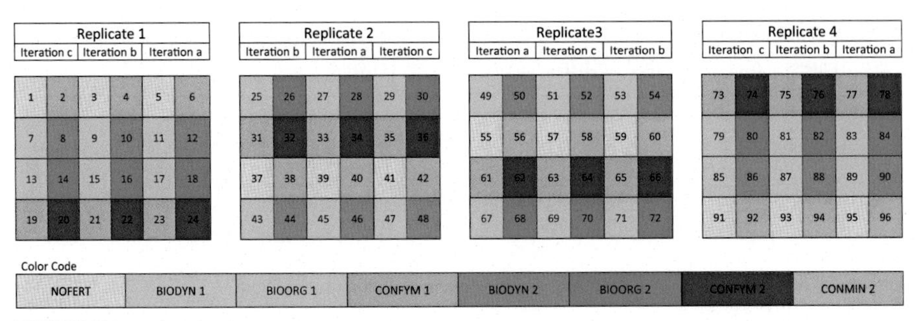

FIGURE 2.1

Spatial arrangement of the experimental plots in the DOK System Comparison Trial with information on blocks (replicates 1–4), systems (color code), iterations (a, b, c), and plot numbers (1–96). Replicated blocks are divided from each other by a 5 m grassland stripe.

corresponds to 0.7 livestock units per hectare. About 1.4 livestock units represent the level of mixed farms in Swiss agriculture. Thus level 2 is the "normal" fertilization level, and level 1 is a reduced level. Nonetheless, 0.7 livestock units may be quite typical for organic farming for other countries, like Germany. Organic Inputs for BIODYN, BIOORG, and CONFYM systems are sourced from local contracted farmers according to the respective label. Notably, biodynamic manure compost preparation requires sophisticated methods and longer periods of processing compared with rotten or stacked manure applied in BIOORG and CONFYM.

Nutrient demands of CONFYM not covered by organic fertilization via manure or slurry are complemented in mineral form up to the limits of Swiss fertilization recommendations (Richner and Sinaj, 2017). In contrast, CONMIN receives solely mineral fertilizers (NPK) up to the limits of these standards. Consequently, nutrient inputs in both conventional systems are considerably higher than in both organic systems. The amount of P and K applied in mineral form in the conventional systems also considers soluble forms of these elements in the soil. Note that nitrogen in manure in CONFYM is not fully accounted for as plant available, and thus total nitrogen input is higher in CONFYM than in CONMIN (Table 2.2).

For the organic fertilization in BIODYN, BIOORG, and CONFYM, the amount of raw material defines the organic fertilization intensities and enters the processing of manure in each system. Processing of raw manure differs in oxygenation status, duration, and handling (Table 2.2). Consequently, the system boundary for organic matter includes the composting process, and BIODYN receives 15% less organic matter via manure than BIOORG and CONFYM because of higher carbon losses during composting. As nutrient analyses of manure and slurry are not always readily available at the time of fertilization, approximate amounts might be given during field management. After nutrient analysis, actual inputs are accounted for and adjusted during the course of each cropping period. Summed up over five crop rotations, organic farming systems are managed less intense and receive 38% less total nitrogen, 71% less mineral nitrogen, 34% less phosphorous, and 29% less

Table 2.2 Mean annual nutrient inputs during the crop rotation 1–5 in comparison to CONFYM2. Treatments with level 1 intensity receive half of organic inputs and chemical fertilizers compared with level 2.

	Total N	Mineral N (NH_4^+, NO_3 in fertilizers)	Total P	Total K	Organic matter
		kg ha^{-1} a^{-1}			
BIODYN 2	99	30	25	172	1900
BIOORG 2	100	32	27	176	2199
CONFYM 2	162	106	39	247	2244
CONMIN	119	119	38	238	–

potassium compared with CONFYM (Table 2.2). While farmyard manure is used as basic fertilization strategy, slurry is used to adjust apparent nutrient demand, mainly N, of the given crop. The effective N input when including biological N fixation through leguminous plants in the rotation is distinctly higher in all systems and is taken into account in N balances. Generally, organic farming systems receive manure and slurry at smaller dosages but more frequent in the rotation crops compared to CONFYM systems.

The 7-year crop rotation with 2 years of grass clover ley is typical for Swiss mixed farming systems combining livestock husbandry and arable cropping. The crop rotation in each 7-year period is modified to best possibly adapt to advances and inplementation within regionally applied agricultural practice (Table 2.3). One example is the implementation of corn and soy cultivation in the fourth crop rotation because of its increased share in Swiss agricultural landscapes and also enhanced use in organic agriculture. However, the crop rotation in the DOK trial is also a compromise between needs and peculiarities of organic and conventional agricultural practice. In reality, organic systems include a higher share of grass clover while conventional systems cultivate more cereals. Changes in crop rotations are discussed in detail in Crop rotation changes in all treatments section, but in general root crops (e.g., potato), grass clover, and cereals (e.g., winter wheat, barley) are part of every crop rotation. Intercropping with legumes during the first three crop rotations and green manures are used to reduce bare fallows in all farming systems. Soil management includes primary tillage with a plow to a depth of 20 cm and rotary harrow and cultivator use for seed bed preparation in all farming systems. However, noninverting soil management practices like hoeing are used more frequently in the organic systems as measure of weed control (Table 2.1). However, mechanical equipment for soil management evolved over time. Conventional and organic systems were managed similarly after the second crop rotation onward (see Section Soil primary tillage harmonization and fertilization adjustments). Pest, disease, and weed control is achieved by chemical crop protections with insecticides, fungicides, and herbicides in both conventional systems. Economic thresholds were respected when these means were applied. Pest and disease controls in NOFERT,

Table 2.3 Changes in the 7-year crop rotation since the start of the DOK field trial.

1. Crop rotation 1978–1984	**2. Crop rotation 1985–1991**	**3. Crop rotation 1992–1998**	**4. Crop rotation 1999–2005**	**5. Crop rotation 2006–2012**	**6. Crop rotation 2013–2019**
Potato *Green manure*	Potato *Green manure*	Potato	Potato	Silage corn	Silage corn *Green manure*
Winter wheat 1 *Intercropping*	Winter wheat 1 *Intercropping*	Winter wheat 1 *Intercropping*	Winter wheat 1 *Green manure*	Winter wheat 1 *Green manure*	Soy
Cabbage	Beetroot	Beetroot	Soy *Green manure*	Soy *Green manure*	Winter wheat 1 *Green manure*
Winter wheat 2	Winter wheat 2	Winter wheat 2	Silage corn	Potato	Potato
Barley	Barley	Grass clover 1	Winter wheat 2	Winter wheat 2	Winter wheat 2
Grass clover 1	Grass clover 1	Grass clover 2	Grass clover 1	Grass clover 1	Grass clover 1
Grass clover 2	Grass clover 2	Grass clover 3	Grass clover 2	Grass clover 2	Grass clover 2

BIODYN, and BIOORG base merely on indirect measures such as crop rotation and the planting of resistant varieties. In BIOORG, copper is used to control late blight of potatoes and organic systems utilize biological pest control measures to regulate Colorado beetles. The crop rotation, which is key for preventive control of pests, diseases, and weed infestation, was matter of change to optimize the agronomic and environmental performance of the systems in the course of the experiment.

Changes of systems and management

The DOK trial was conceptualized as a semistatic experiment. The two pillars that are distinctly different between the farming systems are type and amount of fertilizers and means of plant protection (Table 2.1). There was a constant adjustment and optimization of these management practices in each system, without losing representativeness. In addition, standards of label organizations always were respected. The crop rotation in the DOK trial started with a model rotation typical for the region and was identical in all systems. As management practices evolve over the course of years and decades, the treatments in the DOK trial were faced with changing management practices to ensure their practical relevance. Given the vast range of management options in each farming system, the DOK trial aims at representing standard practices according to the guidelines of the respective labels. Each change in management practice is discussed in the advisory board among scientist and farmers. Most changes become effective at the start of a new crop rotation to ensure stable and comprehensible management. Owing to the design of the field trial, the advisory board is faced with the question whether the systems should reflect current average of regional agricultural practices or define best possible practices. Mimicking current practice results in a monitoring of agricultural performance of high practical relevance. However, following this approach, there is little potential to contribute to the evolution of best management practices with the backup of scientific data. Also in this regard, the management of the DOK trial aims to find balance between these approaches, and, for example, a 7-year crop rotation with 2 years of grass clover can be seen as best management practice for conventional systems.

Although larger changes were implemented in the DOK trial, the requirements of system specific cropping practice were always fulfilled. Retrospectively we identify several major drivers for rotation changes: plant health and yield performance, nutrient cycling, market demand, and new emerging crops. In addition, considerations of an efficient management of the DOK trial with respect to labor were influencing the redesign of the rotation.

System changes

After the first crop rotation, a major change was made with respect to systems on the DOK trial. The system CONMIN replaced an unfertilized control treated with chemical plant protection. In the first rotation, the two unfertilized controls served as a reference for biodynamic and conventional systems. The introduction of a purely mineral fertilization regime in CONMIN added a broader scope to the field trial. At first, CONMIN acts as a second control treatment, which enables the comparison of prevailing farming systems in Switzerland with purely mineral fertilization strategies common in Europe and on a global scale. In Switzerland, purely mineral fertilized cropping systems are still a minority, but they become more widespread in particular in Western Switzerland for short-term economic reasons and as a strategy to reduce labor demand. On the other hand, the CONMIN system operates within the principles of Swiss IP and thus can be considered as a second conventional system. With the implementation of CONMIN, the focus of the DOK trial shifted from a national field trial, investigating the feasibility of biodynamic and organic agricultural practice toward a system comparison trial of broader significance.

Crop rotation changes in all treatments

The first change of the rotation was realized after the first crop rotation when beetroot replaced cabbage. Site conditions requested labor-intensive irrigation measures for cabbage and high pressure of pests, especially *pieris rapae* was receded. Furthermore, cabbage as a transplanted crop was also quite labor intense and on farmers' advice beetroot, another vegetable crop of local importance, replaced cabbage. The advantage was that it was direct seeded without transplanting and there was little disease and pest prevalence. High pressure of a disease led to the omission of barley from the crop rotation. Cultivating three cereal crops within 4 years, as practiced within the first two crop rotations with winter wheat, barley, and winter wheat, led to increased appearance of soil born root and stem diseases, like the fungi *Pseudocercosporella herpotrichoides*. Moreover, the incidence of a weed grass *Apera spica-venti* caused high pressure in the cereals, in particular in treatments with low fertilizer level. Consequently, barley was replaced by a third year of grass clover after the second crop rotation when also organic fertilization intensities were increased to 0.7 and 1.4 LU ha^{-1} $year^{-1}$. This also matched the increased fodder demand following the increase of organic fertilization intensity. From the fourth crop rotation onward, silage corn replaced the third year of grass clover. This change maintained the amount but diversified the source to feed livestock. Inclusion of the high demand crop in the crop rotation also reflected increasing regional importance of silage corn cultivation, mainly driven by conventional farmers. With time, organic farmers increasingly adapted silage corn cultivation at a regional yield gap of ~13% (DOK trial mean yields crop rotation 4–6). Together with silage corn, soybean, another emerging crop was included in the fourth crop rotation of the DOK trial because of soybeans nutritional value and increasing share in regional cultivation. Intensive breeding in Switzerland revealed cold-tolerant soybean varieties adapted to the local site conditions. The legume proved as valuable part of the crop rotation because of its importance as protein source and capability to symbiotic nitrogen fixation. Especially the organic systems benefited from the introduction of soy and the low demands on soil nitrogen status enabled similar yields compared with conventional systems (+0.6% mean yields in organic systems of the DOK, crop rotation 4–6).

From the fourth to the sixth crop rotation, the components of the crop rotation remained stable but still crops were shifted within the sequence of cultivation to deal with pest pressure and crop nutrient supply. After 2 years grass clover, the rotation started with silage corn instead of potatoes. The two and a half years of grass clover provided favorable conditions for beetles of the Elateridae family and their larvae (wire worms) increasingly complicated potato production after termination of grass clover. With the cultivation of silage corn, this pest attack was drastically reduced and silage corn with its high nutrient demand over a long growth period could benefit from high nitrogen fixation and mineralization after grass clover incorporation. Wheat yield after silage corn, in particular in organic systems, was poor because the soil mineral nitrogen stocks were exploited. By changing the position of wheat and soybean, wheat could profit from the fixed nitrogen of the soy, and the risk of *Fusaria* infestation of wheat after silage corn was minimized. Although it seems that for the upcoming crop rotation, a good balance and a stable sequence of elements had been found, the history of crop rotation development shows us that the next change in management is just a matter of time. Agricultural practice is constantly evolving, refining, and redesigning and so are the management practices within the treatments of the DOK trial.

Variety changes in all treatments

An integral part of agricultural practices is the choice of varieties cultivated in the DOK trial. Changes of varieties are exemplified here for wheat. The choice of a wheat variety is a constant issue for the advisory board, but all varieties have to fulfill general requirements: First varieties need to be recommended for organic and conventional farming systems by federal institutions and FiBL. Second, the quality of the wheat needs to be listed among the best performing cultivars for bread wheat with a high baking quality (class I, top). As a third and fourth criteria, the yield performance and the resistance against pests and diseases are taken into account. Consequently, high yields are not the only aim during variety choice, which also reflects the aim of the field trial to act as example for environmental-friendly agricultural practice. Because new varieties constantly emerge because of breeding efforts for higher yields, better resistance, and constant high baking quality, new breads were planted in the course of the DOK trial. The need for change was also provoked by current actual varieties which became susceptible because of loss of resistance when pathogens adapted over time. Although the choice of varieties might differ among the actually practiced organic and conventional farming systems of the region, in the DOK trial the same crop variety was sown in organic and conventional systems. This allowed also for in-depth wheat quality analyses focusing on proteomics and metabolomics. For wheat, almost every crop rotation a new variety was introduced. At first "Sardona" replaced the variety "Probus," which was cultivated in the first crop rotation. "Probus" was tall and not suited to high input because of lodging. In the third crop rotation, the wheat varieties "Ramosa" and "Tamaro" were cultivated. In the fourth crop rotation, only "Tamaro" was cultivated as

"Ramosa" was discarded from the federal list of recommended cultivars, and no seeds were available. In the fifth and sixth crop rotation, the variety "Runal," widespread among conventional practitioners, entered the crop rotation as first winter wheat. To balance between the farming systems, the variety, "Wiwa," bread under organic conditions for low-input systems, was selected as second winter wheat crop. The continuous variety change thus reflects locally applied cropping practice.

Soil primary tillage harmonization and fertilization adjustments

Plowing is used as primary tillage to prepare soils. Within the first two crop rotations, plowing depth in the organic farming systems was 15–20 cm. In contrast, conventional systems including NOFERT were plowed to a depth of 20–25 cm. After adoption of Swiss IP principles in the third crop rotation, plowing depth was leveled in all farming systems to a depth of 20 cm. This was also done as a measure to simplify work and reduce labor costs and to reflect the attempt of Swiss-IP for a more environmental-friendly food production. However, noninverting soil management practices such as harrowing and hoeing are used more frequently in the organic systems as a measure of weed control (Table 2.1). Concomitant with the inclusion of three out of seven fodder crops in the rotation from the third crop rotation onward, the stocking density in the BIODYN, BIOORG, and CONMIN system was increased from 0.6 to 1.2 LU to 0.7 and 1.4 LU for the half- and full-fertilization intensities.

The use of mineral nitrogen sources in CONFYM and CONMIN provides easily available nitrogen sources for plant growth. In the first three crop rotations, ammonium nitrate was used as mineral fertilizer form, but a steady soil acidification was observed likely connected to the use of this N form. Since the fourth crop rotation, calcium ammonium nitrate was used as N fertilizer, which aims at maintaining soil pH at a stable level through the addition of calcium. In addition, lime was applied within the fourth crop rotation in both conventional treatments. In 1999, liming with 0.75 t ha^{-1} CaO equivalents was performed in the treatments CONFYM and CONMIN (Oberholzer et al., 2009). This increased buffer capacity and soil pH in the CONFYM treatment. For CONMIN, another liming treatment with 2 t ha^{-1} CaO equivalents was necessary at the end of the fourth crop rotation in 2005 to prevent further soil acidification. Meanwhile soil pH in BIODYN and BIOORG remained stable, but further decreased in NOFERT. While in CONFYM and CONMIN, liming was performed as exceptional management practice to maintain soil quality indicating by surpassing pH threshold levels, the soil acidification in NOFERT was accepted within the system approach of an unfertilized control treatment.

Plant protection adjustments

The adoption of the principles of "Swiss integrated production" in the CONFYM and CONMIN treatment after the second crop rotation was another major management change implemented in the DOK trial. This followed the foundation of the label IP-Suisse in 1989. By adopting the Swiss IP principles, a clearly defined and regionally applied food production system became the model for the implementation of management practices in the CONFYM treatment with several implications. For example, the amount of herbicides, fungicides, and pesticides declined substantially because of the introduction of economic thresholds for use of chemical plant protection. In addition, some very harmful pesticides, such as Dinoseb with active ingredient Dinoterb to terminate potato growth, were replaced with a more environmental-friendly herbicide. It has also to be considered that the lower amounts of pesticides applied in the conventional systems have to be evaluated in view of ultralow dosage pesticides, which are highly effective. Although the environmental performance of the CONFYM treatment benefited from the adoption of IP-Suisse principles, the amount of plant protection measures in CONFYM and CONMIN still exceeded BIOORG and BIODYN by far. At the same time, fertilization doses in CONFYM were to be adjusted to current soil nutrient contents and to the fertilization guidelines for conventional, which allow for more N per hectare in recent years because of higher yields. The routine mineral N analyses in soil at the beginning of growth in spring allow for a target N application for arable crops in the conventional systems; however, these practices never entered farm practice at large scale in Switzerland.

Conclusions

Owing to further growing concerns on the environmental impact of agricultural practice (Tuomisto et al., 2012; IPBES, 2019) and the need to adapt to projected climate change scenarios (IPCC, 2019), investigating agronomic and environmental performance of organic and conventional farming systems depicts an important societal necessity (Muller et al., 2017). The importance of the DOK long-term trial for this challenging task was highlighted by the Swiss federal authorities by listing the DOK long-term field trial among the Swiss National Research Infrastructure in 2017. Although the field trial was rather designed to investigate agronomic feasibility of organic farming systems, the research focus in the last years shifted toward environment performance of organic and conventional systems. The unique platform of DOK trial was utilized in a series of projects of the Swiss National Science Foundation and the recent European research program Horizon (2020). For the development of new methods, e.g., in the field of metagenomics, proteomics, and remote sensing, the DOK materials are one of the most valuable material sources. After four decades, the field trial still offers a scientific platform to answer current issues in agricultural land use and with every year, the trial becomes even more valuable. Long-term field data are crucial for applied agricultural research and as differences in soil properties emerge, the field trial further offers the opportunity to contribute to basic research on the functionality of organic and conventional farming systems. Given the history of management changes implemented in the DOK long-term field trial to maintain relevance to applied cropping practices, the field trial is well-positioned to contribute to further emerging research questions on arable land use systems.

References

Besson, J.-M., Vogtmann, H., Lehmann, V., Augstburger, F., 1978. DOK: Versuchsplan und erste Ergebnisse eines Projekts zum Vergleich von drei verschiedenen Anbaumethoden. Schweizerische landwirtschaftliche Forschung Band 17, Heft3/4 191–209.

Besson, J.-M., Lehmann, V., Soder, M., 1986. Der DOK Versuch: Ein Projekt zum Vergleich dreier anbaumethoden. Schweizerische Landwirtschaftliche Zeitung. Jahrgang 114 (Nummer 22), 20–28.

Biggs, S.D., 1985. A farming systems approach: some unanswered questions. Agricultural Administration 18, 1–12.

Carson, R., 1962. Silent Spring. Houghton Mifflin, Boston.

Fließbach, A., Oberholzer, H.-R., Gunst, L., Mäder, P., 2007. Soil organic matter and biological soil quality indicators after 21 years of organic and conventional farming. Agriculture, Ecosystems & Environment 118, 273–284.

IPBES, 2019. Summary for policymakers of the global assessment report on biodiversity and ecosystem services of the Intergovernmental Science-Policy Platform on Biodiversity and Ecosystem Services. In: Díaz, S., Settele, J., Brondízio, E.S., Ngo, H.T., Guèze, M., Agard, J., Arneth, A., Balvanera, P., Brauman, K.A., Butchart, S.H.M., Chan, K.M.A., Garibaldi, L.A., Ichii, K., Liu, J., Subramanian, S.M., Midgley, G.F., Miloslavich, P., Molnár, Z., Obura, D., Pfaff, A., Polasky, S., Purvis, A., Razzaque, J., Reyers, B., Roy Chowdhury, R., Shin, Y.J., Visseren-Hamakers, I.J., Willis, K.J., Zayas, C.N. (Eds.), IPBES Secretariat. Bonn, Germany. 56 pages.

IPCC, 2019. Summary for policymakers. In: Shukla, P.R., Skea, J., Calvo Buendia, E., Masson-Delmotte, V., Pörtner, H.-O., Roberts, D.C., Zhai, P., Slade, R., Connors, S., van Diemen, R., Ferrat, M., Haughey, E., Luz, S., Neogi, S., Pathak, M., Petzold, J., Portugal Pereira, J., Vyas, P., Huntley, E., Kissick, K., Belkacemi, M., Malley, J. (Eds.), Climate Change and Land: An IPCC Special Report on Climate Change, Desertification, Land Degradation, Sustainable Land Management, Food Security, and Greenhouse Gas Fluxes in Terrestrial Ecosystems.

Mäder, P., Fliessbach, A., Dubois, D., Gunst, L., Fried, P., Niggli, U., 2002. Soil fertility and biodiversity in organic farming. Science 296, 1694–1697.

Mäder, P., Fliessbach, A., Dubois, D., Gunst, L., Jossi, W., Widmer, F., Oberson, A., Frossard, E., Oehl, F., Wiemken, A., Gattinger, A., Niggli, U., 2005. The DOK Experiment (Switzerland). ISOFAR Scientific Series, No 1: Long Term Experiment in Organic Farming. Verlag Dr. Köster Berlin, pp. 41–58.

Meadows, D.-H., Meadows, D.-L., Randers, J., Behrens, W., 1972. The Limits to Growth. A Report for the Club of Rome's Project for the Predicament of Mankind. Universe Books, New York.

Muller, A., Schader, C., El-Hage Scialabba, N., Brüggemann, J., Isensee, A., Erb, K.-H., Smith, P., Klocke, P., Leiber, F., Stolze, M., Niggli, U., 2017. Strategies for feeding the world more sustainably with organic agriculture. Nature Communications 8 (1), 1290.

Oberholzer, H.-R., Fließbach, A., Mäder, P., Mayer, J., 2009. Einfluss von biologischer und konventioneller Bewirtschaftung auf biologische Bodenqualitätsparameter: Entwicklungen im DOK Langzeitversuch nach pH-Regulierung. Beiträge zur 10. Wissenschaftstagung Ökologischer Landbau.

Reganold, J., Wachter, J., 2016. Organic agriculture in the twenty-first century. Nature Plants 2, 15221.

Richner, W., Sinaj, S., 2017. Grundlagen für die Düngung landwirtschaftlicher Kulturen in der Schweiz (GRUD 2017). Agrarforschung Schweiz 8 (6). Spezialpublikation, 276 S.

Schmid, O., 2007. Development of standards for organic farming. In: Lockeretz, William (Hrsg.) Organic Farming. An International History. CAB International, Wallingford, UK, pp. 152–174 (Kapitel 8).

Seufert, V., Ramankutty, N., 2017. Many shades of gray—the context-dependent performance of organic agriculture. Science Advances 3.

Seufert, V., Ramankutty, N., Foley, J.A., 2012. Comparing the yields of organic and conventional agriculture. Nature 485, 229–232.

Tuomisto, H., Hodge, I., Riordan, P., Macdonald, D., 2012. Does organic farming reduce environmental impacts?—A meta-analysis of European research. J. Environ. Manage. 112, 309–320.

CHAPTER

Pursuing agroecosystem resilience in a long-term Mediterranean agricultural experiment

3

N. Tautges[1], K. Scow[2]

[1]*Chief Scientist, Agricultural Sustainability Institute, Russell Ranch Sustainable Agriculture Facility, University of California Davis, Davis, CA, United States;* [2]*Professor, LAWR, University of California, Davis, CA, United States*

One way of fostering this long view is through 'listening places'—places set aside for patient and oft-repeated measurements, where our observations are melded into those of our predecessors, then handed off as heirlooms to those who follow us. In that way, we bequeath a lengthening legacy—a library expanding with time—from which to read the soil's memory and elicit portents of what is yet to be.

H. Henry Janzen

Introduction

The highly productive and diverse agroecosystems of Mediterranean regions are considered the grocery baskets of the world, producing over 5% of the world's cereals, 7% of pulses, and 9% of vegetables, despite comprising less than 1% of global land area (FAOSTAT, 2019; Rubel and Kottek, 2010). Mediterranean regions are characterized by hot, dry summers and mild, wet winters during which the vast majority of precipitation falls. These conditions create environments with relatively low humidity and disease pressure in the summer, high growing degree days, and a highly controlled water supply, all of which form prime environmental conditions for crop production, especially for vegetables. Irrigation is almost always mandatory in Mediterranean agroecosystems for vegetable and fruit production. Water, nutrient, and carbon fluxes, as well as management practices, tend to differ dramatically from those in humid continental and temperate regions.

Despite their productivity, Mediterranean regions are expected to be more susceptible to rising temperatures and drought associated with impending climate change than temperate agricultural production regions (Muñoz-Rojas et al., 2012; Romanyà and Rovira, 2011). Increased variability of rainfall and severity of drought is expected in Mediterranean cropping systems as climate change progresses

Long-Term Farming Systems Research. https://doi.org/10.1016/B978-0-12-818186-7.00004-7
Copyright © 2020 Elsevier Inc. All rights reserved.

(Iglesias et al., 2012). While irrigated cropland is a small percentage of cropland globally, irrigation is extremely important in Mediterranean systems and for global food supplies, as irrigated agriculture accounts for over 50% of the food consumed globally (Iglesias et al., 2011). Sustainability of irrigation water supplies is particularly vulnerable to climate change (Giorgi and Lionell, 2008). Given this uncertainty in future reliable supplies of irrigation water, which would have ripple-down negative effects on crop productivity and soil health outcomes, the creation of *resilient* agroecosystems that maintain productivity and biological functioning when faced with abiotic and biotic stresses will be essential to securing food production in Mediterranean regions.

Long-term agricultural experiments provide unique and valuable testbeds to explore agricultural resilience, climate smart farming, and both mitigation and adaptation to climate change. However, long-term agricultural research has been neglected in Mediterranean regions relative to the large presence of long-term cropping systems experiments in temperate regions, as evident in the available literature (Diekow et al., 2005; Eivazi et al., 2006; Goulding et al., 2000; Liebman and Davis, 2000; Mäder et al., 2002; Snapp et al., 2010; Teasdale et al., 2007).

The Century Experiment (formerly known as "LTRAS") was established in 1993 in Yolo County at the University of California, Davis's Russell Ranch Sustainable Agriculture Facility (RRSAF) to investigate issues of sustainability confronting Mediterranean agriculture. The overall goal is to quantify the impact of different management practices on resource use efficiency, crop productivity, and soil health changes among irrigated and rainfed cropping systems (Wolf et al., 2018). The original cropping systems established emphasized two crop rotations and treatment structures: (1) processing tomato (*Solanum lycopersicum* L.)—maize (*Zea mays* L.) rotations with varying management systems approaches (conventional, organic, and hybrid) and (2) wheat (*Triticum aestivum* L.)—fallow with an additive treatment structure comparing +/− irrigation, nitrogen (N) fertilizer, and winter cover crops (WCC; Table 3.1). Soils at the site include Yolo silt loam (fine-silty, mixed, nonacid, thermic Typic Xerorthent) and Rincon silty clay loam (fine, montmorillonitic, thermic Mollic Haploxeralf) (Kong et al., 2011).

Over the first 15 years of the RRSAF project, studies have addressed impacts of farming system management on crop yields (Denison et al., 2004), effect of cover crops on cropping systems (McGuire et al., 1998; Hasegawa et al., 1999; Hasegawa and Denison, 2005), impact of management on fruit nutritional quality in tomatoes (Mitchell et al., 2007), and changes in soils and crops with transition into organic management (Martini et al., 2004). Extensive research on soils has included studies of the effects of agricultural management on soil carbon and aggregation (Kong et al., 2005, 2007; Fonte et al., 2007) and soil microbial community composition and function (Kong et al., 2010, 2011; Okano et al., 2004; Lundquist et al., 1999a, 1999b, 1999c). Tillage systems have also been compared in tomato—maize rotations (Kong et al., 2009; Minoshima et al., 2007).

Since the RRSAF project was initiated in 1994, however, there have been significant changes in California agriculture in response to global economic drivers,

Table 3.1 Period 1. Original crop rotations, irrigation, and fertility inputs treatments in the Century Experiment, from 1993 through 2012.

Crop rotation	Irrigation	Fertility source
Tomato–maize	Furrow	Synthetic N fertilizer
WCC[a]/tomato–WCC[a]/maize	Furrow	Synthetic N fertilizer + WCC[a]
WCC[a]/tomato–WCC[a]/maize[b]	Furrow	Poultry manure compost + WCC[a]
Tomato–wheat	Furrow	Synthetic N fertilizer
Wheat–fallow	None	None
Wheat–fallow	None	Synthetic N fertilizer
Wheat–WCC[a]/fallow	None	WCC[a]
Wheat–fallow	Sprinkler	None
Wheat–fallow	Sprinkler	Synthetic N fertilizer
Wheat–WCC[a]/fallow	Sprinkler	WCC[a]

SSDI, *subsurface drip irrigation;* WCC, *winter cover crops.*
[a] *Mix of bell bean (*Vicia faba*), hairy vetch (*Vicia villosa*), and oat (*Avena sativa*).*
[b] *Certified organic system.*

land-use pressures in the state, resource depletion (especially water), and climate change (Wolf et al., 2017). The goal of this paper is to provide an overview of recent research at the RRSAF Century Experiment after 25 years of management. In particular, we will discuss new directions in research on agricultural resilience, climate smart farming, and both mitigation and adaptation to climate change.

Changing times

Original systems: Originally, the Century Experiment was designed around a set of wheat–fallow systems and tomato–maize rotations that received different inputs and management practices. The wheat–fallow treatments established in 1993 reflected agronomic research priorities of the 1990s that often emphasized fallow in semiarid areas and inputs combination comparisons, systems now generally considered to be scientifically reductive and environmentally degradative (Kremen et al., 2012; Tonitto et al., 2006). More innovative were the tomato–maize management systems comparisons, inspired by early systems–based thinking that asked how ecological (e.g., organic) management practices compared to conventional/synthetic chemical inputs-based management that prevailed following the Green Revolution. In addition, the inclusion of a third system—a hybrid conventional + WCC system—was particularly innovative at the time, as ecological versus conventional management comparisons are often based on assumptions of all-or-nothing dichotomies (e.g., nonchemical certified organic agriculture vs. sole reliance on chemicals, no tillage vs. intensive tillage, permanent cover vs. fallow, etc.). Only more recently have hybrid approaches to agricultural management been emphasized (Reganold and Wachter, 2016).

System changes in response to California land-use change, technological innovations, and global change: Agricultural land use has changed dramatically in California's Central Valley since the establishment of the Century Experiment. Progress in surface and groundwater diversion and pumping technologies have expanded the footprint of irrigated agriculture and increased production of irrigated high-value vegetables and perennial tree nuts such as almonds and walnuts.

Rapid expansion of new orchards, facilitated by expansion of high-pressure irrigation technologies, has displaced traditional annual row crop systems once predominant in the valley, particularly wheat production in the foothills of the Coastal and Sierra Nevada mountain ranges. Furthermore, adoption of innovations such as subsurface drip irrigation (SSDI; Fig. 3.1), which can reduce water use by up to 50%, has largely replaced flood/furrow irrigation in annual systems. This trend has been accelerated by a competitive and costly labor market (labor inputs are greater for furrow irrigation than for SSDI) and by state policy and societal pressure to reduce agricultural water use in the state. Wheat acreage that was in rotation with higher value crops has been further reduced as a result of increased conversion to SSDI because drip tends to be damaged during lack of use with a wheat crop, greatly reducing its life span from a potential of up to 10 years. Although these changes in land use and opportunities for new high-value crops have decreased local grower interest in wheat–fallow systems and furrow irrigation, wheat–fallow systems are important cropping systems both nationally and globally in semiarid climates. RRSAF provides important California data for these larger networks and will continue to contribute to the body of research around these systems. Furthermore,

FIGURE 3.1

Flood irrigation of furrows, between crop beds (left) and subsurface drip irrigation (right), with drip tape buried at 25 cm.

we have undertaken studies with crops like intermediate wheatgrass (Kernza), a new perennial cereal, to push innovation in nonirrigated semiarid agroecosystems.

To address these political, social, and economic changes across the California agricultural landscape and to evolve with new agro-scientific paradigms, in 2013, RRSAF researchers converted furrow irrigation systems to SSDI for all tomato and maize crops and replaced the irrigated wheat—WCC/fallow system (Table 3.1) with a "Native Grass" system to represent restored and managed California rangeland ecotypes with native grass spp. (Table 3.2). Also in 2013, we added a 3-year alfalfa—tomato—maize system that integrates annual and perennial crops, grain and forage production, and furrow and SSDI approaches (Table 3.2). Lastly, a conventional tomato—maize system with integrated use of both synthetic N fertilizers and poultry manure compost was added in 2018. This change was motivated by scientific and management questions about differences between conventional versus organic versus hybrid systems, where only the WCC were present in hybrid management, and the effect of compost could not be isolated from the old treatment structure (Tables 3.1 and 3.2).

Making changes to long-term cropping systems experiments inevitably provokes debate and criticism among researchers and the public, as changes to systems or practices are often interpreted as "undermining" previous multiyear research investments and endangering the robustness of data and results. These conflicts often

Table 3.2 Period 2. Crop rotations, irrigation, and fertility inputs treatments in the Century Experiment, from 2013 through present.

Crop rotation	Irrigation	Fertility source
Tomato—maize	*SSDI*	Synthetic N fertilizer
WCC[a]/tomato—WCC[a]/ maize	*SSDI*	Synthetic N fertilizer + WCC[a]
WCC[a]/tomato—WCC[a]/ maize[b]	*SSDI*	Poultry manure compost + WCC[a]
Tomato—maize[c]	*SSDI*	Synthetic N fertilizer + poultry manure compost[c]
Alfalfa—alfalfa—alfalfa—tomato—maize—tomato	*Furrow/ SSDI*	*Synthetic N fertilizer*
Native grass[d]	*None*	*None*
Wheat—fallow	None	None
Wheat—fallow	None	Synthetic N fertilizer
Wheat—WCC[a]/fallow	None	WCC[a]
Wheat—fallow	Sprinkler	None
Wheat—fallow	Sprinkler	Synthetic N fertilizer

Significant changes were made to treatment structures in 2013; changes made from periods 1 to 2 are italicized. SSDI, *subsurface drip irrigation;* WCC, *winter cover crops.*
[a] *Mix of bell bean (*Vicia faba*), hairy vetch (*Vicia villosa*), and pea (*Pisum sativum*).*
[b] *Certified organic system.*
[c] *Conventional tomato—maize + composted poultry manure system added in 2018.*
[d] *Mix of* Elymus multisetus, Melica californica, Nassella cernua, Nassella pulchra, Poa secunda, *and* Vulpia microstachys.

occur when crop managers consider changing crop varieties, for example. Changing crop varieties can affect the yield potential of the system, confounding analyses of internal and external factors' effects (e.g., tillage treatments, fertility inputs, precipitation levels, etc.) on crop yields. However, agriculture does not stand still but continually adapts to challenges from new pests, diseases, and technologies through the development of new crop varieties, agrochemicals, and management practices such as irrigation. Similarly, to maintain relevance to farmers and policymakers, the farming systems of long-term experiments need to keep up to reflect significant changes in management. While changing cropping systems and experimental treatments should always be undertaken only after deliberate thought and planning, we argue that for long-term cropping systems to be resilient, they must remain relevant, in the face of short-term and variable funding structures increasingly being used to sustain these long-term sites that demand consistent funding.

Furthermore, while consistency of management is important for quality of data and results, it is equally important that the long-term experimental facility be a place where researchers can innovate and propel agroecological research forward. To achieve this, systems and management practices must continually evolve along with new technologies and paradigms throughout the life of the experiment or the experiment ends up becoming more of a museum than a place for research. However, flexibility of treatment and cropping systems structures must be balanced by consistency in design and implementation to preserve continuity and control environmental variables over time.

Innovating for resilience

A major focus at RRSAF has been on organic amendments—looking at both mechanisms and practices—in Mediterranean row crops and how they impact indicators of agroecosystem sustainability. Three of our tomato–maize rotations differed substantially in amounts of carbon added as residues and amendments: (1) conventional management with synthetic chemical fertilizers, winter fallow, and pesticides, which added 4.3 Mg C ha^{-1} yr^{-1}; (2) certified organic with composted poultry manure (fertility input) and WCC, which added 7.3 Mg C ha^{-1} yr^{-1}; and (3) a hybrid system with synthetic fertilizers, pesticides, and WCC, which added 5.1 Mg C ha^{-1} yr^{-1}. We have compared productivity outcomes among these systems for 24 years. The incorporation of organic amendments, in the form of compost in the certified organic system, significantly increased the yield stability and decreased variation in tomato yields in the long term, compared to conventional, external inputs-based systems (Fig. 3.2).

Notably, the addition of organic matter inputs to soil in the form of composted poultry manure and WCC increased tomato yield stability, whereas the addition of WCC to a conventional system dependent on synthetic fertilizer inputs did not change yield stability over time. Lower variation in yields suggests that the organic system's productivity is less responsive to year-to-year environmental variability and more resilient to climate- and pest-induced stresses. Resilience to heat and

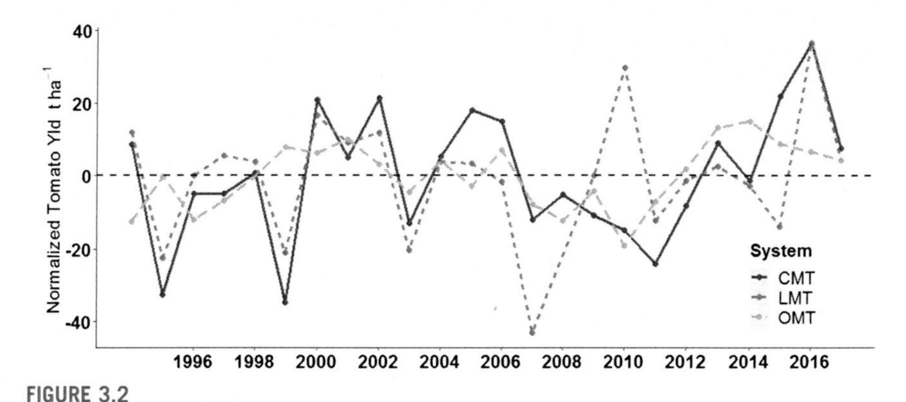

FIGURE 3.2

Variance of tomato yields in the conventional, conventional + winter cover crops (WCC), and certified organic compost + WCC systems, depicted as normalized fruit yields relative to 30-year mean tomato yields, at the Russell Ranch site.

drought in the organic system was particularly evident in the years 2006 and 2015, years when summer heatwaves in conjunction with drought severely impacted tomato crops across the state. However, organic tomato yield variance was 50% and 63% lower, respectively, than in conventional systems, including the conventional + WCC system. Annual compost applications to the organic system seemed to be driving changes in system-wide resilience, possibly via enhanced water availability. The organic system had significantly greater water infiltration rates than the conventional systems with and without WCC after 25 years of inputs, suggesting that water storage belowground was enhanced by compost inputs in particular. However, the actual mechanisms of agroecosystem resilience and interactions between resilience and soil health indicators are complex and require more study (Fig. 3.3).

Soil health for agroecosystem resilience: irrigated crops

A major emphasis of the Century Experiment has been on soil health. The incorporation of agroecological practices into tomato–maize rotations, such as winter cover cropping and compost application, has driven major changes in many soil physical and biological properties and associated soil health indicators among the RRSAF systems. A soil health assessment in collaboration with the United States Department of Agriculture-Natural Resources Conservation Service (USDA-NRCS) was performed in 2017 after 24 years of consistent conventional, hybrid, and organic management with WCC and compost. We used the Cornell Soil Health Assessment to compare soil health indicators in the different management systems (Idowu et al., 2009). The Cornell Assessment was developed for Northeastern United States where climate, soils, and agricultural practices differ substantially from Mediterranean ecosystems. Therefore, we compared the indicator values from the assessment but did not use the individual ratings and overall quality score of soil health. Consistent

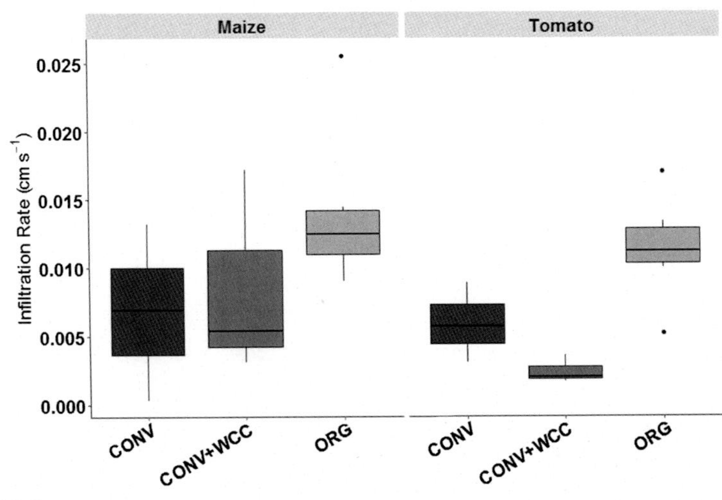

FIGURE 3.3

Water infiltration rate, measured with a double-ring infiltrometer post harvest of maize and tomato crops in 2018, among conventional (CONV), conventional with winter cover crops (CONV + WCC), and organic with compost and cover crops (ORG) management in tomato–maize systems.

Data courtesy of Daoyuan Wang, UC Davis.

changes in soil health properties over time were evident in many of the chemical and biological soil health indicators which tended to have higher values in the organic compost + WCC system, intermediate in the conventional + WCC, and lowest in the conventional system with no organic matter inputs besides crop residues. The indicators reflecting physical properties of soils were generally less sensitive to management than the biological and chemical indicators.

Soil organic matter was affected most by compost inputs and showed the greatest increases in the organic compost + WCC system, followed by the conventional + WCC, and lowest in the conventional system (Fig. 3.4A). Soil microbial activity and populations, which are governed in large part by carbon availability, increased significantly with the addition of organic matter inputs, for both WCC alone (conventional + WCC) and compost + WCC in the organic system, which had the greatest microbial biomarkers and activity (Fig. 3C and D). Aggregate stability, a physical property of soil positively correlated with water infiltration and resistance to crusting (Abiven et al., 2009), increased with additional organic matter inputs to soil over time. However, the addition of compost to WCC in the organic system did not increase aggregate stability above what was measured in the conventional + WCC system (Fig. 3.4B).

In general, sustainable, soil health-building practices like cover cropping and compost amendments had markedly larger impacts on the soil biological indicators than physical and chemical indicators. This suggests that a lack of carbon inputs in

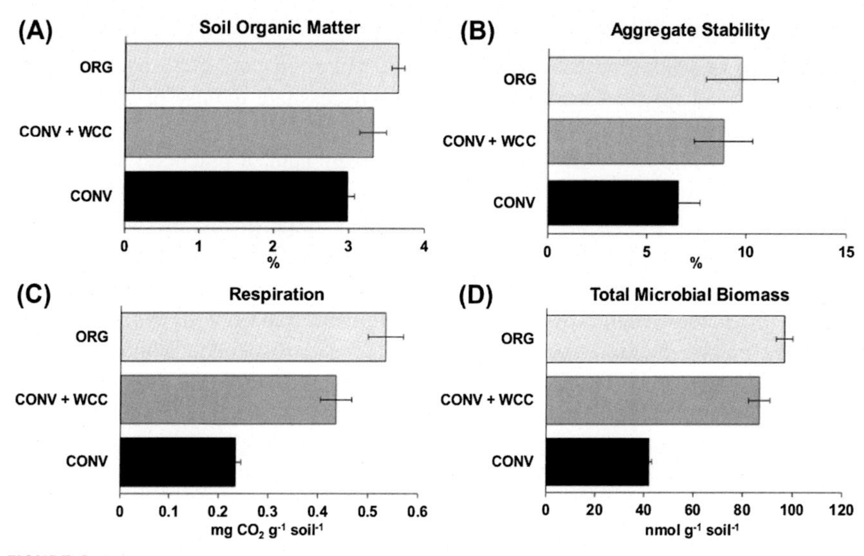

FIGURE 3.4

Soil health indicators in the top 0–15 cm after 24 years of conventional (CONV), conventional with winter cover crops (CONV + WCC), and organic with compost and cover crops (ORG) management in tomato–maize systems.

Data were obtained from the Cornell Soil Health Assessment, courtesy of J. Kucera-Moore, NRCS.

terms of quantity or quality is a limiting factor for soil microbial growth and functioning in annual conventional row crop systems characterized by winter fallowing and synthetic chemical inputs. Microbial activity is essential for many soil functions and services, thus stimulation of microbial growth and activity via cover cropping and compost inputs likely supports greater conversion of organic matter inputs into stable soil organic matter, particularly through promotion of soil aggregation (Six et al., 2004).

Long-term cropping systems sites like RRSAF are invaluable for demonstrating the potential for changes in soil health indicators that can occur with organic matter–building practices, which can be substantial (as observed in this study) in the long term but are difficult to detect in the short term. Indeed, the inability to observe tangible changes in soil properties in the first or second year of cover crop and compost use adoption is a barrier to adoption of these practices by growers (Carolan, 2006; Rodriguez et al., 2008). For growers and other stakeholders, long-term trials with side-by-side comparisons of different farming practices can help transcend temporal barriers to observing soil change processes that occur on multiyear to decadal scales, like soil organic matter building. Rather than having to wait 5–10 years for management investments to manifest, long-term trials can provide growers a look into the future to observe how decades of sustainable management practices can affect soil attributes.

Soil health for agroecosystem resilience: nonirrigated crops

Many crops in Mediterranean regions, particularly grains, are traditionally rainfed. Development of resilient agroecosystems may necessitate the use of crops that do not require irrigation water to produce food, like wheat. When water is scarce or unavailable, many growers are forced to fallow their land, a practice that can have negative environmental effects such as increased soil erosion and loss of soil carbon (Campbell et al., 1997). At the RRSAF Century Experiment, the wheat–fallow systems are sometimes perceived as a relic from before the days of irrigated agriculture. However, these systems are indeed important areas of research as climate change progresses. Understanding the environmental trade-offs in net primary productivity, carbon fixation and storage, and water use associated with nonirrigated grain crops and fallowing, and their relevant management practices will be essential for identifying adaptation strategies in Mediterranean agroecosystems. Furthermore, the addition of the wheat–fallow + WCC system at RRSAF allows a novel exploration of the effects of incorporating a soil health-building practice into low-intensity, low-input semiarid systems.

Soil health indicators after 24 years of wheat–fallow cropping systems inputs comparisons suggest that N inputs via either synthetic chemical fertilizers or biological N fixation (by leguminous WCC) increase soil microbial activity and biomass, compared to a rainfed wheat–fallow system with no inputs which provides a "control" treatment and a system representing pre–Green Revolution agriculture (Table 3.3). This suggests that modern N management, using either chemical N inputs or leguminous crops in rotation with grain crops, may have increased soil microbial growth and activity relative to historical levels on croplands managed without leguminous ley phases and/or manure inputs. Supplemental winter irrigation of the wheat–fallow that received no N increased both microbial activity and biomass (Table 3.3), suggesting that even in N-limited systems, greater soil water can increase soil biological functions. However, in the wheat–fallow + N systems, supplemental irrigation did not increase microbial activity and biomass; and the incorporation of carbon with N in the form of WCC did not increase microbial

Table 3.3 Biological soil health indicators in the top 0–15 cm after 24 years of wheat–fallow (W–F) systems +/– nitrogen (N) chemical fertilizer inputs and winter cover crops (WCC).

	Respiration		Total microbial biomass	
Cropping system	**mg CO_2 g^{-1} $soil^{-1}$**	**Se**	**nmol g^{-1} $soil^{-1}$**	**Se**
Rainfed W–F control	0.226	0.008	48.5	0.691
Rainfed W–F + N	0.280	0.015	68.7	5.93
Rainfed W–F + WCC	0.273	0.006	73.6	3.17
Irrigated W–F	0.244	0.009	53.9	1.88
Irrigated W–F + N	0.290	0.015	61.3	3.51

Data courtesy of J. Kucera-Moore, NRCS.

indicators to a greater extent than N fertilizer. These results suggest that N is limiting to wheat growth and possibly soil C sequestration in Mediterranean wheat—fallow systems. Therefore, in these low-input wheat—fallow systems that might be grown on marginal lands, addition of N fertilizer would be a higher priority than supplemental irrigation for enhancing biological indicators of soil health.

Soil carbon sequestration for climate change adaptation and resilience

Threats to Mediterranean agroecosystems from climate-induced changes in temperature and moisture regimes could create a positive feedback loop by decreasing net primary productivity (Potter et al., 2012) and increasing soil carbon decomposition rates (Davidson and Janssens, 2006) in these regions. Mediterranean soils have inherently low soil carbon contents (Aguilera et al., 2013), which presents both challenges and opportunities for climate change adaptation. Sustaining soil fertility, health, and plant productivity in the face of rising temperatures may be endangered by low soil organic matter levels; however, Mediterranean agricultural soils have been identified as a large potential repository for carbon sequestration, as these soils have carbon levels far below their saturation points (Romanyà and Rovira, 2011; Muñoz-Rojas et al., 2012).

Global initiatives to increase carbon sequestration in agricultural and working lands, like the "4 per mille" initiative (French Ministry of Agriculture, 2015), aim to offset a significant portion of global CO_2 emissions by reallocating atmospheric C to long-term nonreactive forms of soil carbon. Regardless of whether soil carbon levels are increased enough to offset a significant portion of emissions (Poulton et al., 2018), efforts to increase soil carbon levels, both in terms of concentrations and stocks (quantity), could contribute to agroecosystem resilience across small and large scales. Given the long time scales of carbon building processes in soils, it has been difficult to identify processes and magnitudes of soil carbon change in small-plot, short-term plots typical of agricultural field research. Long-term research sites and networks of these sites should be leading the effort to identify carbon sequestration potentials and practices that maximize stored carbon across different soil classes and climatic regions, to drive decision-making models and policies that incentivize carbon building and storage in soils.

The state of California, in one of the first government-led programs in the United States to pay growers for carbon credits, enacted an incentives program called the "Healthy Soils Initiative" to increase soil health- and carbon building practices on farmland, by awarding growers grants to implement these practices on their farms (California Department of Food and Agriculture, 2018). Management practices eligible for incentives were generally based on the US National Resource Conservation Service (NRCS) recommendations for building soil health, which include reduced tillage, cover crops, and residue retention, to which compost application was also added (California Department of Food and Agriculture, 2018).

At the RRSAF, we have been monitoring soil carbon levels for over 20 years among the cropping systems of the Century Experiment, providing a comparison of the effects of WCC alone and WCC + poultry manure compost (in the hybrid and organic systems, respectively; Table 3.2) on soil carbon changes over time in tomato–maize rotations. Soil carbon data measured in years 0–19 (1994–2012) to a 2-m depth indicated that addition of WCC alone did not increase soil carbon levels relative to a conventional system with winter fallow, when taking into account changes in bulk density and carbon fluxes throughout the 2-m soil profile. However, the organic system with compost in addition to WCC carbon inputs increased soil carbon by 12% from baseline to year 19.

Our results conflict somewhat with the conclusions of a metaanalysis by Poeplau and Don (2015), which found that the addition of WCC alone to an annual crop rotation increased soil carbon stocks by 0.32 Mg C ha^{-1} yr^{-1}, although their analysis was restricted to the top 30 cm in which we also observed increases in C in our conventional + WCC plots. However, at depths below the top 30 cm—the usual focus of agricultural studies—soil carbon dynamics are quite different, as was observed in RRSAF soil carbon analysis. It is essential to note here that individual agricultural management practices, within both scientific and policy arenas, are often discussed as being uniformly applicable across climatic regions (especially within the United States), despite drastically varying winter–summer temperature and precipitation cycles. Therefore, the effect of new management practices on a desired outcome, such as soil carbon building, must be evaluated within multiple climatic zones worldwide. While WCC may increase soil carbon in temperate regions of the world, differing soil–water–microbial dynamics in winter versus summer among temperate and Mediterranean regions could lead to different outcomes from WCC in a variety of variables, including soil carbon, nutrient cycling, and microbial diversity. Region-specific recommendations for management practices will be essential for agroecosystem adaptation to climate change going forward.

Regardless of the limitations of WCC alone to raise soil carbon, the large benefits of composted manure, in addition to WCC cultivation, on carbon sequestration were clear. Given that the incorporation of both compost and WCC had benefits on soil health indicators (Fig. 3.3) as well as soil carbon sequestration, RRSAF long-term research clearly supports increasing the use of compost in Mediterranean agroecosystems for a range of purposes, both environmental and agronomic. The organic system's approach of combining more than one soil health-building practice, using both WCC and compost, speaks to a diversified approach that leads to *multifunctionality* of the agroecosystem, to a greater extent than implementation of just one short-term practice, like cover cropping. Often, if one agricultural management practice alone does not substantially satisfy an economic or environmental goal, there are many who argue for the disuse of that one practice. These arguments have become ubiquitous in arguments over many aspects of agroecosystem management, including tillage (Govaerts et al., 2009; Huggins and Reganold, 2008; Rochette, 2008), soil carbon sequestration (Minasny et al., 2017; De Vries, 2017), and organic agriculture (Mäder et al., 2002; Reganold, 1988; Trewavas, 2001).

As approaches to strategic system design hopefully become less dichotomous going forward, as is argued for by some (Pimentel et al., 2005; Reganold and Wachter, 2016; Searchinger et al., 2018) and exemplified by new experimental approaches such as the alfalfa–tomato–maize rotation at the RRSAF, then creating agroecosystems that aspire toward multifunctional and resilience outcomes may, on the other end, employ multifaceted means to achieve those goals.

Engagement with farmers and policymakers—role of long-term research experiments

Another potential role for long-term research experiments is to provide a physical meeting place and intellectual hub where science-based, mechanistic research can connect with knowledge derived from the experiences of farmers and other practitioners. Given the many challenges facing state and national governments, and the planet, around the sustainability of agricultural production, it is more urgent than ever that research works to solve on-the-ground problems. To ensure relevance, researchers need to have opportunities to hear concerns and questions identified by farmers and policymakers.

The RRSAF Facility, through its Century Experiment and outreach events, builds and strengthens relationships among researchers, farmers, and policymakers, particularly in California. Our approach at RRSAF is to create a variety of opportunities for connection. These include outreach activities which range from more traditional field days for researchers to share their latest results to special workshops that focus on targeted issues, such as the microbiology of soil health. We have hosted policymakers from the State—California Department of Food and Agriculture; California Energy Commission; Air Resources Board—providing a working farm for them to observe irrigation equipment, agricultural machinery, and management practices, as well as learn about specific research protocols that provide monitoring data for climate change policies. RRSAF is also part of California's Healthy Soil Program, which incentivizes the management of farms and ranches specifically for carbon sequestration and augmentation of cobenefits such as increased water-holding capacity and soil fertility, funding projects including direct farmer incentives and farmer-to-farmer demonstration networks to increase adoption of soil management practices that sequester carbon and increase the resilience of agricultural systems.

Research on agroecosystem resilience is not just about biology, chemistry, and physics, and not just about technological advances. If the research is to result in change, the social dimension must also be considered. Science produces lots of seemingly sound ideas that are never adopted for a variety of reasons. For example, a technology may be inappropriate for the targeted users because of site-specific conditions, high costs, or conflicts with local agricultural traditions. Farmers and other stakeholders must be involved, ideally from the start, as discoveries are made and implemented, to guide future directions and evaluate impacts. Long-term agricultural experiments can provide common ground on which to build these important relationships. From these relationships, researchers investigating mechanisms driving responses to global change can be more successful in developing more realistic decision support models that incorporate economic, agronomic, ecological, and social trade-offs. These tools can then provide support for both growers and policymakers to make sound management decisions in the face of increasing climate uncertainty.

Acknowledgments

This work is/was supported by the USDA National Institute of Food and Agriculture, Hatch Project CA-2122-H, and the USDA National Institute of Food and Agriculture Grant #11925159. Any opinions, findings, conclusions, or recommendations expressed in this publication are those of the author(s) and do not necessarily reflect the view of the National Institute of Food and Agriculture (NIFA) or the United States Department of Agriculture (USDA).

References

Abiven, S., Menasseri, S., Chenu, C., 2009. The effects of organic inputs over time on soil aggregate stability—a literature analysis. Soil Biology and Biochemistry 47, 1–12.

California Department of Food and Agriculture, 2018. The Office of Environmental Farming & Innovation: Healthy Soils Program. https://www.cdfa.ca.gov/oefi/healthysoils/.

Campbell, C.A., Janzen, H.H., Juma, N.G., 1997. Case studies of soil quality in the Canadian Prairies: long-term field experiments. In: Gregorich, E.G., Carter, M.R. (Eds.), Soil Quality for Crop Production. Elsevier Science Publishers, Amsterdam, The Netherlands.

Carolan, M., 2006. Do you see what I see? Examining the epistemic barriers to sustainable agriculture. Rural Sociology 71, 232–260.

Davidson, E.A., Janssens, I.A., 2006. Temperature sensitivity of soil carbon decomposition and feedbacks to climate change. Nature 440, 165–173.

Denison, R.F., Bryant, D.C., Kearney, T.E., 2004. Crop yields over the first nine years of LTRAS, a long-term comparison of field crop systems in a Mediterranean climate. Field Crops Research 86, 267–277.

De Vries, W., 2017. Soil carbon 4 per mille: a good initiative but let's manage not only the soil but also the expectations: comment on Minasny et al. Geoderma 292, 111–112.

Diekow, J., Mielniczuk, J., Knicker, H., Bayer, C., Dick, D.P., Kögel-Knabner, I., 2005. Carbon and nitrogen stocks in physical fractions of a subtropical Acrisol as influenced by long-term no-till cropping systems and N fertilization. Plant and Soil 268, 319–328.

Eivazi, F., Bayan, M.R., Schmidt, K., 2006. Select soil enzyme activities in the historic Sanborn Field as affected by long-term cropping systems. Communications in Soil Science and Plant Analysis 34, 2259–2275.

Fonte, S.J., Kong, A.Y., van Kessel, C., Hendrix, P.F., Six, J., 2007. Influence of earthworm activity on aggregate-associated carbon and nitrogen dynamics differs with agroecosystem management. Soil Biology and Biochemistry 39, 1014–1022.

Food and Agriculture Organization of the United Nations, 2019. FAOSTAT Statistics Database. http://www.fao.org/faostat/en/#data.

Giorgi, F., Lionell, P., 2008. Climate change projections for the Mediterranean region. Global Plant Change 63, 90–104.

Goulding, K.W.T., Poulton, P.R., Webster, C.P., Howe, M.T., 2000. Nitrate leaching from the Broadbalk Wheat Experiment, Rothamsted, UK, as influenced by fertilizer and manure inputs and the weather. Soil Use & Management 16, 244–250.

Govaerts, B., Verhulst, N., Castellanos-Navarrete, A., Sayre, K.D., Dixon, J., Dendooven, L., 2009. Conservation agriculture and soil carbon sequestration: between myth and farmer reality. Critical Reviews in Plant Science 28, 97–122.

Hasegawa, H., Labavitch, J.M., McGuire, A.M., Bryant, D.C., Denison, R.F., 1999. Testing CERES model predictions of N release from legume cover crop residue. Field Crops Research 63, 255–267.

Hasegawa, H., Denison, R.F., 2005. Model predictions of winter rainfall effects on N dynamics of winter wheat rotation following legume cover crop or fallow. Field Crops Research 91, 251–261.

Huggins, D.R., Reganold, J.P., 2008. No-till: the quiet revolution. Scientific American 299, 70–77.

Idowu, O.J., van Es, H.M., Abawi, G.S., Wolfe, D.W., Schindelbeck, R.R., Moebius-Clune, B.N., Gugino, B.K., 2009. Use of an integrative soil health test for evaluation of soil management impacts. Renewable Agriculture and Food Systems 24, 214–224.

Iglesias, A., Mougou, R., Moneo, M., Quiroga, S., 2011. Towards adaptation of agriculture to climate change in the Mediterranean. Regional Environmental Change 11, S159–S166.

Iglesias, A., Quiroga, S., Moneo, M., Garrote, L., 2012. From climate change impacts to the development of adaptation strategies: challenges for agriculture in Europe. Climatic Change 112, 143–168.

Kong, A.Y., Six, J., Bryant, D.C., Denison, R.F., Van Kessel, C., 2005. The relationship between carbon input, aggregation, and soil organic carbon stabilization in sustainable cropping systems. Soil Science Society of America Journal 69, 1078–1085.

Kong, A.Y., Fonte, S.J., Van Kessel, C., Six, J., 2007. Soil aggregates control N cycling efficiency in long-term conventional and alternative cropping systems. Nutrient Cycling in Agroecosystems 79, 45–58.

Kong, A.Y., Fonte, S.J., van Kessel, C., Six, J., 2009. Transitioning from standard to minimum tillage: trade-offs between soil organic matter stabilization, nitrous oxide emissions, and N availability in irrigated cropping systems. Soil and Tillage Research 104, 256–262.

Kong, A.Y., Hristova, K., Scow, K.M., Six, J., 2010. Impacts of different N management regimes on nitrifier and denitrifier communities and N cycling in soil microenvironments. Soil Biology and Biochemistry 42, 1523–1533.

Kong, A.Y., Scow, K.M., Córdova-Kreylos, A.L., Holmes, W.E., Six, J., 2011. Microbial community composition and carbon cycling within soil microenvironments of conventional, low-input, and organic cropping systems. Soil Biology and Biochemistry 43, 20–30.

Kremen, C., Iles, A., Bacon, C.M., 2012. Diversified farming systems: an agroecological, systems-based alternative to modern industrial agriculture. Ecology and Society 17, 44–63.

Liebman, M., Davis, A.S., 2000. Integration of soil, crop and weed management in low-external-input farming systems. Weed Research 40, 27–47.

Lundquist, E.J., Jackson, L.E., Scow, K.M., 1999a. Wet–dry cycles affect dissolved organic carbon in two California agricultural soils. Soil Biology and Biochemistry 31, 1031–1038.

Lundquist, E.J., Scow, K.M., Jackson, L.E., Uesugi, S.L., Johnson, C.R., 1999b. Rapid response of soil microbial communities from conventional, low input, and organic farming systems to a wet/dry cycle. Soil Biology and Biochemistry 31, 1661–1675.

Lundquist, E.J., Jackson, L.E., Scow, K.M., Hsu, C., 1999c. Changes in microbial biomass and community composition, and soil carbon and nitrogen pools after incorporation of rye into three California agricultural soils. Soil Biology and Biochemistry 31, 221–236.

Mäder, P., Fliessbach, A., Dubois, D., Gunst, L., Fried, P., Niggli, U., 2002. Soil fertility and biodiversity in organic farming. Science 296, 1694–1697.

Martini, E.A., Buyer, J.S., Bryant, D.C., Hartz, T.K., Denison, R.F., 2004. Yield increases during the organic transition: improving soil quality or increasing experience? Field Crops Research 86, 255–266.

McGuire, A.M., Bryant, D.C., Denison, R.F., 1998. Wheat yields, nitrogen uptake, and soil moisture following winter legume cover crop vs. fallow. Agronomy Journal 90, 404–410.

Minasny, B., Malone, B.P., McBratney, A.B., Angers, D.A., Arrouays, D., Chambers, A., Chaplot, V., Chen, Z.S., Cheng, K., Das, B.S., Field, D.J., Gimona, A., Hedley, C.B., Hong, S.Y., Mandal, B., Marchant, B.P., Martin, M., McConkey, B.G., Winowiecki, L., 2017. Soil carbon 4 per mille. Geoderma 292, 59–86.

Minoshima, H., Jackson, L.E., Cavagnaro, T.R., Sánchez-Moreno, S., Ferris, H., Temple, S.R., Goyal, S., Mitchell, J.P., 2007. Soil food webs and carbon dynamics in response to conservation tillage in California. Soil Science Society of America Journal 71, 952–963.

Mitchell, A.E., Hong, Y.J., Koh, E., Barrett, D.M., Bryant, D.E., Denison, R.F., Kaffka, S., 2007. Ten-year comparison of the influence of organic and conventional crop management practices on the content of flavonoids in tomatoes. Journal of Agricultural and Food Chemistry 55, 6154–6159.

Muñoz-Rojas, M., Jordán, A., Zavala, L.M., De la Rosa, D., Abd-Elmabod, S.K., Anaya-Romero, M., 2012. Organic carbon stocks in Mediterranean soil types under different land uses (Southern Spain). Solid Earth 3, 375–386.

Okano, Y., Hristova, K.R., Leutenegger, C.M., Jackson, L.E., Denison, R.F., Gebreyesus, B., Lebauer, D., Scow, K.M., 2004. Application of real-time PCR to study effects of ammonium on population size of ammonia-oxidizing bacteria in soil. Applied and Environmental Microbiology 70, 1008–1016.

Pimentel, D., Hepperly, P., Hanson, J., Douds, D., Seidel, R., 2005. Environmental, energetic, and economic comparisons of organic and conventional farming systems. BioScience 55, 573–582.

Poeplau, C., Don, A., 2015. Carbon sequestration in agricultural soils via cultivation of cover crops—a meta-analysis. Agriculture, Ecosystems & Environment 200, 33–41.

Potter, C., Klooster, S., Genovese, V., 2012. Net primary production of terrestrial ecosystems from 2000 to 2009. Climate Change 115, 365–378.

Reganold, J.P., 1988. Comparison of soil properties as influenced by organic and conventional farming systems. American Journal of Alternative Agriculture 3, 144–155.

Reganold, J.P., Wachter, J.M., 2016. Organic agriculture in the twenty-first century. Nature Plants 2, 1–8.

Rochette, P., 2008. No-till only increases N_2O emissions in poorly-aerated soils. Soil and Tillage Research 101, 97–100.

Rodriguez, J.M., Molnar, J.J., Fazio, R.A., Sydnor, E., Lowe, M.J., 2008. Barriers to adoption of sustainable agriculture practices: change agent perspectives. Renewable Agriculture and Food Systems 24, 60–71.

Romanyà, J., Rovira, P., 2011. An appraisal of soil organic C content in Mediterranean agricultural soils. Soil Use & Management 27, 321–332.

Rubel, F., Kottek, M., 2010. Observed and projected climate shifts 1901–2100 depicted by world maps of the Köppen-Geiger climate classification. Meteorologische Zeitschrift 19, 135–141.

Searchinger, T.D., Wirsenius, S., Beringer, T., Dumas, P., 2018. Assessing the efficiency of changes in land use for mitigating climate change. Nature 564, 249–253.

Six, J., Bossuyt, H., Degryze, S., Denef, K., 2004. A history of research on the link between (micro)aggregates, soil biota, and soil organic matter dynamics. Soil and Tillage Research 79, 7–31.

Snapp, S.S., Gentry, L.E., Harwood, R., 2010. Management intensity—not biodiversity—the driver of ecosystem services in a long-term row crop experiment. Agriculture, Ecosystems & Environment 138, 242–248.

Teasdale, J.R., Coffman, C.B., Mangum, R.W., 2007. Potential long-term benefits of no-tillage and organic cropping systems for grain production and soil improvement. Agronomy Journal 99, 1297–1305.

Tonitto, C., David, M.B., Drinkwater, L.E., 2006. Replacing bare fallows with cover crops in fertilizer-intensive cropping systems: a meta-analysis of crop yield and N dynamics. Agriculture, Ecosystems & Environment 112, 58–72.

Trewavas, A., 2001. Urban myths of organic farming. Nature 410, 409–410.

Wolf, K., Herrera, I., Tomich, T., Scow, K., 2017. Long-term agricultural experiments inform the development of climate-smart agricultural practices. California Agriculture 71, 120–124.

Wolf, K.M., Torbert, E.E., Bryant, D., Burger, M., Denison, R.F., Herrera, I., Hopmans, J., Horwath, W., Kaffka, S., Kong, A.Y., Norris, R.F., 2018. The century experiment: the first twenty years of UC Davis' Mediterranean agroecological experiment. Ecology 99, 503.

CHAPTER

Challenges of maintaining relevance to current agricultural issues in a long-term cropping establishment experiment in Canterbury, New Zealand

4

Tim Brooker, Nick Poole, Richard Chynoweth, Abie Horrocks, Phil Rolston, Nick Pyke

Foundation for Arable Research, Christchurch, New Zealand

Introduction

This chapter is a discussion of the Chertsey Establishment Trial (CET), an experiment that was established in Canterbury, New Zealand, in 2003 to compare establishment methodology of dryland and irrigated crops. Crop production and soil quality measurements were collected throughout the first 15 years and some of these results are presented here. Challenges unique to a long-term experiment are discussed, including maintaining relevance to current farming issues, human error, and how changes have been implemented while maintaining the historic experimental treatments. A major challenge that was discovered in February 2018 was that large differences in chemical soil fertility have developed in this experiment, which may have impacted crop production of the treatments. The probable causes of these differences and the future plan for this experiment are discussed.

Annual crop production in New Zealand

New Zealand is a relatively young country where widespread cultivation of agricultural land for crop or pasture production began during the period between 1860 and 1910 (White, 1999). Canterbury is still the major arable production region of New Zealand, largely because of favorable topography, a climate consisting of long sunshine hours and cool nights which provide ideal conditions for grain fill and harvest of grain and seed crops. World record grain yields for both wheat and barley have

Long-Term Farming Systems Research. https://doi.org/10.1016/B978-0-12-818186-7.00005-9
Copyright © 2020 Elsevier Inc. All rights reserved.

been achieved by Canterbury farmers: 16.8 t ha^{-1} of winter wheat in 2017 (NZ Herald, 2017) and 13.8 t ha^{-1} of barley in 2015 (Stuff, 2015). The arrival of European immigrants in the mid-19th century resulted in organized settlement of Canterbury, and a mixed arable–livestock system of farming developed soon after, where a range of annual grain or seed crops were grown for 2–5 years, followed by a period of permanent pasture for approximately the same period of time. This was often in conjunction with forage crops grazed in situ to finish sheep or cattle for slaughter.

Since the 1990s, large areas of farmland have been converted to dairy production with the proliferation of both well and surface water irrigation on the Canterbury plains. Pastoral dairy farming in New Zealand has well-established principles and the basic system can be transferred successfully to almost any farm. Although astute business operators can be very reactive to market signals and adjust the proportion of land area for each enterprise accordingly, ease of management, consistent cash flow, and greater stability of returns have attracted many mixed cropping farmers toward dairy production with the availability of water for irrigation and reliable pasture growth.

The increase in dairy production has provided a new market for locally produced grain as dairy farmers often supplement the pastoral diet with grain. Many mixed cropping farmers now grow forage crops to overwinter dairy cows on their farms under contract, rather than the traditional finishing of sheep or beef cattle. Other farmers have intensified crop production and reduced or eliminated the livestock grazing enterprise altogether, often because of the desire to simplify management or to avoid damage to soil by livestock grazing during periods of wet weather.

Mixed crop farming in Canterbury is among the most diverse and complicated farming systems in the world. It requires the skill set to integrate grazing livestock and agronomic knowledge covering a large range of crops, including cereal, pulse, forage, forage legume, and grass seed and vegetable seed production species requiring both spring and autumn establishment.

Tillage and crop establishment

Historically, cultivation was practiced mainly to control undesirable weeds and pests (McKenzie et al., 1999). The sowing of seeds directly into the soil with no prior cultivation (no-till) did not become a realistic proposition for commercial-scale agriculture until the release of herbicides with no residual activity (Hampton et al., 1999). However, many seed drills have been designed to operate in a cultivated seedbed, where the previous crop residue is either buried by a plow (inversion tillage) or incorporated with surface cultivation (minimum tillage) to maximize speed of breakdown.

The trend toward a more intensive cropping rotation and cultivation associated with this has a greater risk of degrading soils through the depletion of soil organic carbon (C) (Müller et al., 2018). A reduction of tillage intensity (inversion > minimum tillage > no-till) at the establishment of each crop in the rotation can minimize unnecessary degradation of quality indicators such as aggregate stability, erosion potential, and water holding capacity, which are all closely associated with soil C.

Changing establishment methodology of crops often requires substantial investment in new machinery by the farm business. Reduced tillage intensity provides the opportunity to minimize establishment costs by reducing the number of machinery passes and decreasing the timeframe for crop establishment. Greater areas may be sown on time or between periods of poor weather. While these benefits of reducing tillage at crop establishment are generally well understood, farmers are reluctant to invest in such machinery without evidence that yield, and consequently income, will not be compromised.

The Chertsey Establishment Trial

The CET was established in 2003 by the Foundation for Arable Research (FAR) near Ashburton, Canterbury, New Zealand. The initial rationale for the experiment was to provide arable farmers with confidence that a progressive shift toward a reduction in cultivation intensity would not penalize crop production under an intensive annual cropping rotation. It was not designed as a farming system comparison.

The CET was established at FAR's Chertsey Research Site, where a large number of agronomic trials and demonstrations are conducted each year on grain and seed crops. This site had been farmed under a mixed cropping system using intensive cultivation for more than 20 years before the establishment of this experiment. The soil is an Umbrisol, classified under the New Zealand Soil Classification as a Chertsey silt loam (Cox, 1978), with 40–60 cm of topsoil over gravels, holding approximately 100 mm of plant-available water. Annual rainfall is approximately 660 mm year^{-1} evenly distributed throughout the year. Irrigated treatments aim to remove soil moisture deficits and usually receive 250–300 mm of applied irrigation to replace evapotranspiration. Dryland crops can experience moisture stress anytime from September until harvest in the January to March period, depending on rainfall and growth stage of the crop.

Treatments and trial design

The CET was initially set up with six establishment treatments (Table 4.1) consisting of two inversion (plowing) treatments, followed by either one or two passes with a disc-type cultivator; two minimum tillage treatments, involving two passes with a disc-type cultivator operating either only in the surface layer, or deeper with the assistance of rippers to remedy compaction; and two no-till treatments where the crop was sown with no prior cultivation by either Great Plains triple-disc drill from 2003 to 2013 and a John Deere double-disc seed drill from 2014 onward, or a Cross-slot seed drill.

Table 4.1 Original and revised cultivation treatments in the Chertsey Establishment Trial.

Original establishment treatments (2003–2017)				Revised establishment treatments (2017–current)			
Treatment	**Category**	**Cultivation**	**Seed drill**	**Treatment**	**Category**	**Cultivation**	**Seed drill**
1	Inversion	Plow, disc roller	Disc drill	1	Inversion	Plow, Maxi till, Cambridge roll	Disc drill
2	Minimum tillage	Disc roller ×2	Disc drill	2	Minimum tillage	Disc roller x2	Disc drill
3	No-till	Nil	Disc drill	3	No-till	Nil	Disc drill
4	Inversion	Plow, disc roller ×2	Disc drill	1	Inversion	Plow, Maxi till, Cambridge roll	Disc drill
5	Minimum tillage	Disc roller with rippers ×2	Disc drill	2	Minimum tillage	Disc roller x2	Disc drill
6	No-till	Nil	Cross-slot	3	No-till	Nil	Disc drill

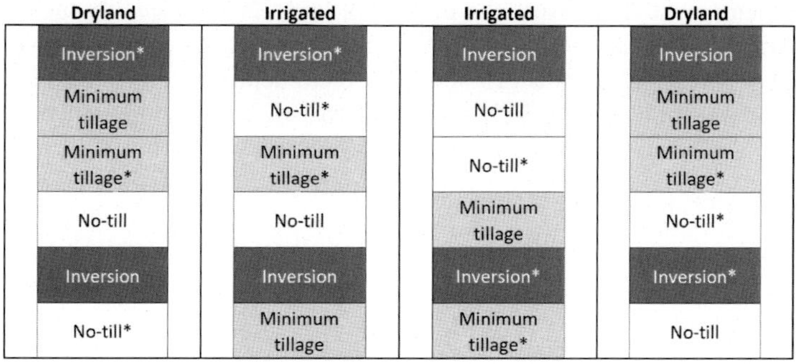

Dryland	Irrigated	Irrigated	Dryland
Inversion*	Inversion*	Inversion	Inversion
Minimum tillage	No-till*	No-till	Minimum tillage
Minimum tillage*	Minimum tillage*	No-till*	Minimum tillage*
No-till	No-till	Minimum tillage	No-till*
Inversion	Inversion	Inversion*	Inversion*
No-till*	Minimum tillage	Minimum tillage*	No-till

FIGURE 4.1

Experimental layout of the Chertsey Establishment Trial. Subplots with asterisks were altered in September 2017.

The experiment is a split-plot design, with irrigation treatments (irrigated and dryland) making up main plots and establishment method as subplots (Fig. 4.1). Owing to constraints of working with farm scale machinery in limited space, the experiment design could not be truly randomized; however, physical examination and electromagnetic conductivity mapping showed no gradient for major soil characteristics across the experimental site. The irrigated plots were positioned adjacent to each other as this aligned with irrigation runs for the rest of the experimental site. The plowed subplots were also kept in line for ease of machinery operation, running perpendicular to the irrigator. Subplots were set at a width of 12 m as this is divisible by 3, 4, and 6, the typical widths for agricultural implements. Buffer areas between the ends of plots give enough room to turn a tractor and implement around without having to traverse other plots.

From September 2017, the six cultivation treatments were reduced to three, by amalgamating the inversion, minimum tillage, and no-till treatments which increased the number of replicates from two to four for each treatment. The section on changes to the trial design to stay relevant will cover the rationale for these changes in detail.

Measurements

Grain or seed yield was measured at each harvest with a small plot combine harvester. Other agronomic measurements made as required include establishment counts of the crop, dry matter cuts to quantify forage production, and weed counts. In addition, soil quality characteristics have been measured to quantify any changes in soil health as a consequence of the establishment and irrigation treatments. Soil quality characteristics were measured in autumn 2003 as a baseline and subsequently in autumn of 2004–09, 2015, 2017, and 2018. The measurements have included penetration resistance, aggregate stability, structural condition scores, earthworm populations, and total soil N and C. From 2003 to 2009, soil C and N measurements were taken from the 0–7.5 cm layer. No soil quality measurements were taken between 2010 and 2014. When the monitoring was resumed in 2015, sampling was carried out at 0–15 and 15–30 cm depth increments. In 2018, sampling was carried out at 0–7.5, 7.5–15, and 15–30 cm depth increments.

Trial management

Crop input management was the same across all treatments until 2018 including crop protection and nutrition inputs and timings. Inputs were usually nonlimiting and based on the treatments with the highest potential for fertilizer and those with the greatest weed, pest, or disease pressure for crop protection. The only management difference among dryland and irrigated treatments was the harvest date, as dryland crops typically reached maturity earlier than the irrigated equivalent.

Thus, the only variables were irrigation and the establishment technique, which ensured that the effects of other management decisions on crop production were eliminated.

Learnings from the first 15 years of the Chertsey Establishment Trial

Crop production

Grain or seed yield at each harvest was measured with a small plot harvester. Cumulative yield of each tillage by irrigation treatment from 2007 to 2018 harvests is presented in Fig. 4.2. The irrigated treatments have cumulatively yielded 50.2 t ha^{-1} of grain or seed over this time, 62% greater ($P < .01$) than the yield of the dryland establishment treatments (31 t ha^{-1}). There was no difference among establishment techniques under the same irrigation treatment or an interaction. From these data, no establishment methodology has shown a clear yield advantage, although no-till establishment under dryland did trend toward a greater yield occasionally.

No harvest data are presented for 2013 as this was the second year of ryegrass seed crop that had not been terminated and consequently none of the establishment treatments were imposed in 2012, simulating a 2-year pasture phase in the rotation.

Soil quality

Soil carbon

Continual cultivation of the soil results in a decline in soil C, due to increased aeration stimulating microbial activity and the oxidation of organic matter (McLaren and Cameron, 1996). Many physical quality attributes of soil such as aggregate stability, structural condition, and water holding capacity are strongly correlated with the level of C present. Soil carbon was monitored in the CET from 2003 to 2009 at a depth of 0–7.5 cm. Soil carbon was lowest ($P < .01$) in both inversion treatments

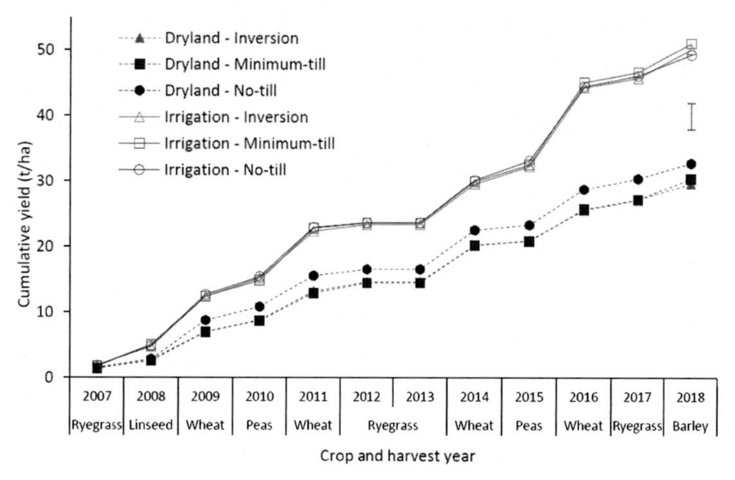

FIGURE 4.2

Cumulative seed and grain yield of crops established with inversion tillage, minimum tillage, and no-till under dryland and irrigation in Canterbury, New Zealand, for 2007–18. Error bar represents the LSD (5%) for all interactions for cumulative yield in 2018 only.

in 2009 at 2.37% and was 2.6% for both the minimum tillage and no-till treatments, supporting that continual inversion tillage was reducing C in the 0–7.5 cm soil layer. No soil C monitoring occurred in 2010–14. In 2015 and 2018, soil C was measured to 30 cm, which found that the previous measurements of only the 0–7.5 cm layer were misleading. There was no difference in soil C in the 0–30 cm layer between any of the establishment treatments in either 2015 or 2018 (Fig. 4.3).

The inversion tillage treatment was burying crop residue to a depth of approximately 20–25 cm, below the depth of sampling for the years 2003–09. Minimum tillage incorporates residue nearer the surface, predominantly in the top 10 cm of soil whereas no-till results in virtually no incorporation of residue. It is important that measurements are taken to below the depth of tillage to understand whether C is being retained at depth or oxidized.

From 2018, sampling has been conducted at 0–7.5, 7.5–15, and 15–30 cm depths to better understand the relationship between establishment method and soil C levels at these depths (Fig. 4.4). No-till treatments had the greatest ($P < .01$) level of C (27.1 t C ha^{-1}) in the 0–7.5 cm fraction, compared with 24.7 t C ha^{-1} for minimum tillage and 21.7 t C ha^{-1} for the inversion tillage treatments, while irrigation had no effect. A recent study by Müller et al. (2018) using intact samples taken from the CET found a reduction ($P < .01$) in soil C of 19% after 13 years when comparing inversion tillage with no-till. In this study, irrigation was quoted to have decreased soil C by 6% ($P = .124$). It was concluded that despite increased biomass production under irrigation, this was broken down faster by increased levels of mineralization (Dersch and Böhm, 2001).

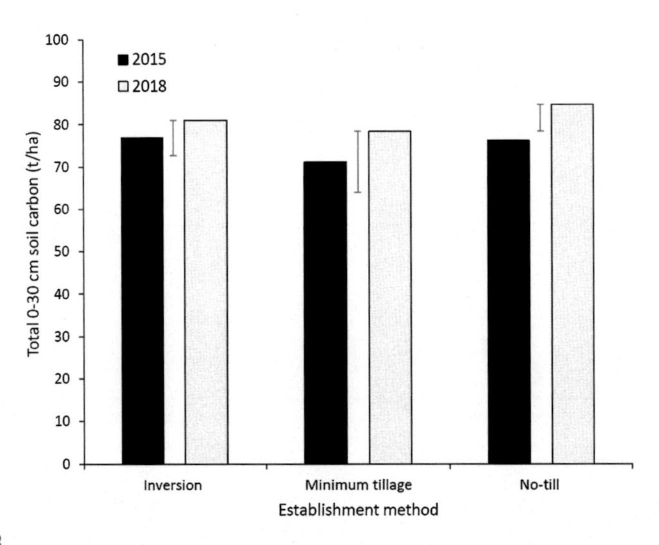

FIGURE 4.3

Total 0–30 cm soil carbon (t ha^{-1}) for years 2015 and 2018 following 11 continuous years of inversion, minimum tillage, and no-till crop establishment in the Chertsey Establishment Trial, Canterbury, New Zealand. Error bars represent LSD 5% between years.

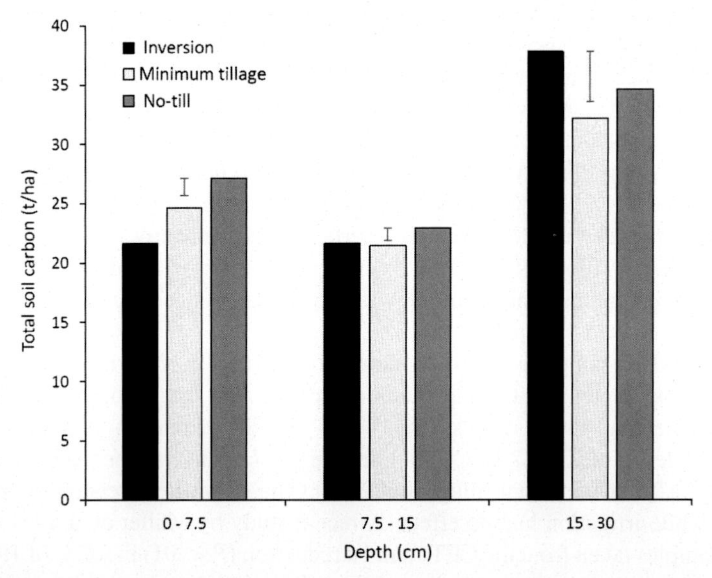

FIGURE 4.4

Soil carbon (t ha^{-1}) in the 0–7.5, 7.5–15, and 15–30 cm fractions for inversion, minimum tillage, and no-till establishment treatments in the Chertsey Establishment Trial, Canterbury, New Zealand. Error bars represent LSD 0.05 between establishment treatments.

Aggregate stability

While no difference in total soil C to 30 cm has been found from the different establishment treatments, the accumulation of greater levels of C in the 0–7.5 cm layer of soil in the no-till treatments has affected physical characteristics such as aggregate stability and structural condition of the soil.

Aggregate stability is a measure of the soil's resistance to structural breakdown. It was assessed using a wet-sieving methodology and expressed as mean weight diameter (MWD). It is very responsive to restorative and degradative phases of the cropping rotation and has shown some clear differences between treatments when other measurements have not. Aggregate stability is a useful indicator of the susceptibility of the soil to erosion, particularly by water movement following periods of heavy rainfall (Barthes and Roose, 2002). In 2008, aggregate stability showed no difference between establishment treatments following 2 years of ryegrass (Fig. 4.5). No-till has shown consistent aggregate stability between 2008 and 2018, ranging from 1.62 to 1.76 mm MWD over this time. Inversion tillage treatments were consistently lower from 2009 to 2018, ranging between 1.42 mm to as low as 0.95 mm MWD, compared with minimum tillage which ranged from 1.41 to 1.83 mm MWD. It is unfortunate that aggregate stability assessment was not conducted in years 2010–2014 as this may have provided further insight. Aggregate stability was 1.07 mm MWD in 2003 at the beginning of this experiment. Although there are many years where measurements were not taken, both no-till and minimum tillage have improved aggregate stability from the initial measurements.

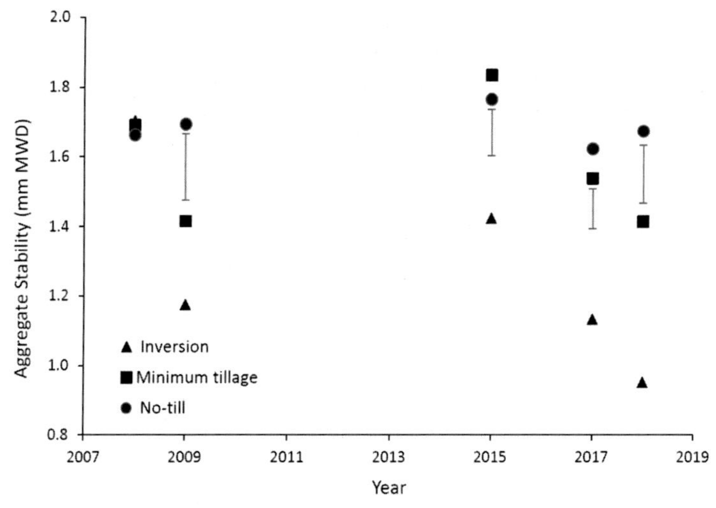

FIGURE 4.5

Aggregate stability (mean weight diameter, mm) of 0–7.5 cm layer of soil of the three establishment treatments in the Chertsey Establishment Trial. Error bars represent LSD (5%) between inversion, minimum tillage, and no-till crop establishment following 2 years of pasture which was terminated in 2008.

Inversion tillage was the most detrimental to aggregate stability; however, this was rapidly improved following 2 years in pasture illustrating the importance of restorative phases in cropping rotations.

Challenges encountered in the Chertsey Establishment Trial

Maintaining the relevance to current farming issues

After 16 years of harvest data collection, no establishment method has demonstrated a consistent yield advantage over the other. While this does not represent a true system comparison between farms, it has to some extent answered the original question—after 16 years, reduced tillage has not been found to suppress crop seed or grain yield.

The CET has received criticism from farmers, in particular those operating conservation no-tillage systems. The argument presented is that changing a farm to no-till requires a rethink of the entire crop rotation, not just the use of a different machine at the establishment. The no-till plots have experienced poor establishment on occasion, particularly of herbage seed crops with a small seed size where excessive crop residue on the soil surface hindered emergence of the following crop. Where burying or incorporation of residues occurred in inversion or minimum tillage, this barrier was removed. In a true no-till system, a farmer would not attempt to establish a ryegrass seed crop following a 12 t ha^{-1} wheat crop. No-till generally lacks popularity in New Zealand's cropping regions where high yields are achieved and consequently there are large amounts of crop residues that can make establishment of the following crop unreliable. Pest pressure at establishment, particularly from slugs, has tended to be greater in no-till. On occasion, the management of this may not have been as timely as it should be, which once again highlights the lack of a whole-system approach.

Stubble burning remains an important tool for many arable farmers in New Zealand. Stubble burning is used to manage weeds, pests, and residue before the establishment of small seeded crops. In practice, burning of crop residues allows for timely establishment of these crops with less cultivation passes. Burning of stubble receives significant opposition from the urban population during the harvest period due to the smoke nuisance. Most farmers recognize the importance of using burning responsibly and only when strictly necessary, to minimize the offense to others. Consequently there has been a 50% reduction in cereal stubble area burnt between 2006 and 2016 (Poole, 2017). Over the same period of time, there has been a 23% increase in noninversion (minimum tillage) crop establishment; however, plowing after pasture remains the dominant establishment technique, with 50% of surveyed farmers using inversion cultivation at this point in the rotation. For some farmers, particularly those that have small seeded crops that may struggle to establish without appropriate management of the previous crop's residue, the loss burning may reverse the trend of reducing establishment intensity until new equipment and knowledge is acquired. Farmers cannot afford the greater risk of failure that no-till establishment of these crops can pose following high-yielding cereal crops in particular.

In general, the establishment methodology used depends on numerous factors including

- what crop is being established,
- the previous crop,
- whether residues are burnt, retained, or baled and removed from the field,
- what weed species are present and chemical control options in the following crop,
- pest or disease pressure,
- whether the soil has been damaged by livestock grazing, particularly on high-yielding winter feed crops, and
- whether compaction needs to be remedied or the surface needs to be releveled.

New Zealand farmers often have a range of cultivation equipment and seed drills that can work in cultivated or uncultivated soil or have access to contractors with these implements. Where there is an opportunity to reduce establishment costs, farmers will capitalize on this, but will use a greater intensity of cultivation when and if they feel obliged.

Due to the nature and original goals of the CET, some compromises were made and not every different farm system could be represented. Overall rotation sequence and residue management were kept consistent between treatments within a growing season; however, inputs to these crops may have varied in timing or possibly quantity to best represent farming under that scenario. The variation of inputs between treatments, if any, must not be pronounced and therefore less likely to cause any long-term unintended consequences that diminish the relevance of the experiment.

Changes to trial design to stay relevant

The six original establishment treatments were originally designed to give varying degrees of soil disturbance to understand the effect of subsequent passes on crop establishment and soil quality. In 2017, discussion with farmers, consultants, and soil scientists distinguished that on-farm tillage was categorized by whether a plow was used, a noninversion tillage regime had been implemented, or no-till occurred for establishment. In general, unpublished data that had been collected from farmers' fields around soil quality supported these categories. Thus, the treatment structure was changed to combine the two plowing, two noninversion cultivation and two direct drilling treatments, respectively, to give three distinct and easily definable treatments: inversion, minimum tillage, and no-till. Inversion now represents the traditional establishment methodology; the field is plowed, followed by a pass with a Maxi till (S-tine cultivator with cage roller) and finally a consolidation pass with a Cambridge roller before sowing. Minimum tillage represents the majority of farmers who have reduced passes and only cultivate the surface layer with two passes of a disc roller combination implement. No-till represents establishment with no prior cultivation and minimal soil disturbance with a double-disc seed drill. The John Deere 750A was retained as the drill used to sow the no-till treatments. The John Deere drill is also used to sow the cultivated treatments, which is logistically much easier than coordinating two contractors to plant on the same day. This alteration to the establishment treatments has also increased the replication for each treatment from two to four, thereby increasing the power of statistical analysis for measurements from the experiment.

Other challenges

A human error mistake was made when plowing in 2006 during which the wrong strip was plowed by the contractor. Three years of cultivation history for these plots was lost. For most of the data around crop productivity, the trial is considered to have started in 2006. Soil quality data from these plots in the 3 years following this error were omitted when calculating long-term trends. In 2016, the contractor accidently plowed one 3-m wide strip through two of the no-till plots. Fortunately, there was still 9 m of both plots that was not cultivated, but this has reduced the area within these plots by one quarter. These mistakes have not just affected 1 year of data, but have reset long-term trends in soil quality.

Staff turnover within the organization has resulted in people with different research ideas and interests being involved, and as a result the experiment has gone through periods of neglect and renewed interest. While the establishment and irrigation treatments have been administered and combinable yield results have always been collected, other measurements and monitoring have been quite sporadic. Data management has also varied considerably and record-keeping has been in varying degrees of detail, which has resulted in some trends or anomalies in data being difficult to explain.

Managing soil fertility uniformity in the Chertsey Establishment Trial

An analysis of chemical soil fertility for each of the treatments was conducted before the establishment of red clover in 2018. The standard 0–15 cm soil core was split into a 0–5 cm sample and a 5–15 cm sample to investigate potential nutrient stratification. The results from this sampling not only revealed expected nutrient stratification, but some major unexpected differences in soil fertility that have resulted from standardized fertilizer inputs across the whole trial (Table 4.2). To minimize financial cost, the soil from each replicate of each treatment was composited and analyzed as one sample.

Soil pH

Soil pH is analyzed in New Zealand using a water dilution (1:2.1 ratio). Suitable pH for cropping in Canterbury ranges from approximately 5.5–6.5, with most farmers targeting a minimum of 5.8. The level of soil acidity in the CET had reached levels

Table 4.2 Soil pH in the Chertsey Establishment Trial in February 2018 following 15 years of uniform crop inputs.

	Dryland			Irrigated		
Depth (cm)	**Inversion tillage**	**Minimum tillage**	**No-till**	**Inversion tillage**	**Minimum tillage**	**No-till**
0–5	5.1	5.1	4.7	5.9	5.5	5.2
5–15	5.6	5.6	5.6	6.0	5.9	5.7
0–15	5.4	5.4	5.3	6.0	5.8	5.5

that were unrealistic in an arable situation and limitations on crop production cannot be discounted because of the unevenness of background soil acidity that had developed between treatments.

Acidification was much greater in all dryland establishment treatments compared with the irrigated plots, both in the 0–5 cm and to 0–15 cm samples. The level of acidification was greatest in the 0–5 cm soil layer of dryland no-till treatments where pH was as low as 4.7; however, the irrigated no-till treatment also had a very low pH of 5.2 (Table 4.2).

The blanket application of N fertilizers to the whole experiment is largely responsible for the development of greater soil acidity in the dryland treatments. Rates of fertilizer N application were calculated to the yield potential of the irrigated treatments, with excess N on the dryland treatments considered nonlimiting to crop production. Dryland treatments consistently yielded half that of the irrigated due to crops experiencing water stress. Consequently not all of the available N in the soil was used by the crop. Excess rainfall over the winter period results in drainage of water through the soil profile, causing nitrate leaching and associated acidification. Aside from the unwanted environmental effects of excess N being lost to groundwater, low pH of soil can lead to high levels of aluminum, which is toxic to plant root development (McLaren and Cameron, 1996). Extreme soil acidity was likely a contributing factor of poor establishment of small seeded crops in no-till treatments, contemporarily to the residue and pest problems associated with previous crops.

The results from soil testing were received in the autumn of 2018, a few weeks before the planned establishment of the red clover seed crop. No records of liming of the trial could be found since the experiment began in 2003, but it was confirmed that there was no liming within the last 6 years. An application of 4 t ha^{-1} of lime was made to the whole trial, immediately before the establishment treatments being administered.

Comprehensive soil chemical analysis of each individual plot was repeated in April 2018, 9 weeks after the application of lime and following red clover establishment. The depths of sampling were altered to 0–7.5 cm, 7.5–15 cm, and 15–30 cm, in line with the depths chosen for other soil physical analyses. Following the lime

Table 4.3 Soil pH in the Chertsey Establishment Trial in April 2018 following lime application in February 2018.

	Dryland			Irrigated		
Depth (cm)	**Inversion tillage**	**Minimum tillage**	**No-till**	**Inversion tillage**	**Minimum tillage**	**No-till**
0–7.5	5.6	6.2	5.6	6.0	6.5	6.1
7.5–15	5.8	5.8	5.6	6.2	5.9	5.8
15–30	5.8	6.1	5.7	6.2	6.1	5.9
0–15	5.7	6.0	5.6	6.1	6.2	5.9
0–30	5.7	6.0	5.7	6.1	6.1	5.9

application, the pH of the top 7.5 cm of soil has been raised to a minimum of 5.6 for all treatments (Table 4.3). While this has not yet reached the targeted value of 6.0, it is less likely to have adverse effects on crop production. Future liming will now be targeted to individual treatments to realign them at a pH of 6.0, otherwise the pH of irrigated treatments risks being risen beyond 6.5. Response of soil pH to liming is gradual, and this will be most apparent in the no-till treatments, as there is not the mechanical disturbance to mix the soil that the cultivated treatments receive.

Farmers who currently establish crops with no-tillage should consider a greater frequency of smaller maintenance lime applications or at very least monitor soil pH closely. Sampling with a New Zealand standard 15 cm soil corer may not accurately represent the level of acidification at the depth of seed placement.

Phosphorus

The Olsen P analysis method of plant-available P is widely accepted in New Zealand agriculture. No-till showed an accumulation of P in the surface layer with an Olsen P of 38 mg L^{-1} and 35 mg L^{-1} for dryland and irrigated, respectively, declining to 12 mg L^{-1} and 9 mg L^{-1} at 15–30 cm (Table 4.4). This trend was slightly less

Table 4.4 Soil Olsen P (mg L^{-1}) in the Chertsey Establishment Trial in April 2018 following 15 years of uniform application.

	Dryland			Irrigated		
Depth (cm)	**Inversion tillage**	**Minimum tillage**	**No-till**	**Inversion tillage**	**Minimum tillage**	**No-till**
0–7.5	22	27	38	15	27	35
7.5–15	21	16	18	16	20	15
15–30	20	9	12	14	10	9
0–15	22	22	28	15	23	25
0–30	21	15	20	15	16	17

pronounced in the minimum tillage treatment where the Olsen P was 27 mg L^{-1} in the surface layer for both irrigated and dryland, which decreased to 9 mg L^{-1} and 10 mg L^{-1} for dryland and irrigated at 15–30 cm depth. The inversion treatment had a relatively even distribution of P throughout the soil profile, with an Olsen P of 21 mg L^{-1} for dryland and 15 mg L^{-1} for irrigated from the entire 0–30 cm profile.

Stratification of P was more apparent in the no-till treatments under both irrigated and dryland; however, further investigation of plant uptake of P and subsequent return to soil via residue incorporation in the different establishment treatments is required to understand what processes are occurring to affect the Olsen P values in the CET.

The varying values of Olsen P between the dryland and irrigated treatments is because of variable removal of nutrients in the crop product. Irrigated plots have consistently yielded approximately double that of dryland ones, and hence nutrient removal in the harvested product would be of the same proportion. Not only is the grain or seed removed, but also straw has been baled and removed following certain crops, as would occur on farm.

Of concern is the level of these Olsen P values for the seed production of some species. Excessive P fertility can be detrimental to the seed yield of most perennial legumes as it encourages vegetative growth, which can suppress flowering. Olsen P soil test results of less than 15 mg L^{-1} are ideal for white clover seed production (Clifford, 1985). It is assumed that red clover seed production is maximized under similar conditions. The current range of Olsen P values varies as a result of the treatments in this trial, thus it cannot be concluded that difference in seed yield of clover between treatments was influenced by the treatments or the background fertility differences that had emerged.

Future of the Chertsey Establishment Trial

The strategy of keeping all inputs other than the establishment machinery passes, and the same was an effort to remove variables which influence crop yield. However, uniform management has resulted in unintended consequences that are compromising the integrity and results from this experiment. For long-term research of this nature, a "farm system" approach needs to be considered; in this instance, farmers would apply base fertilizer and lime reactionary to soil test values, opposed to the previous rationale where inputs were applied to the treatment with the highest demand, or in some cases an average of the whole trial area. There needs to be recognition of trends that develop as a result of establishment treatments and these need to be measured. While individual plot application is more expensive and time-consuming, the differences in soil fertility that have emerged must be realigned for this experiment to continue to provide relevant results.

A long-term experiment needs to stay relevant to the needs of the stakeholders. The CET is funded via a levy taken from the sale of arable crops in New Zealand.

Initially the question was whether farmers could successfully implement reduced cultivation practices on their farms without suffering yield penalty. To date, the results show that establishment methodology does not affect yield in most instances; however, in a dryland scenario, the no-till had trended toward a yield advantage. The advantage may have been compromised by having soil fertility outside the optimum range in more recent years. It would be of interest in the future to investigate the response of crop production between the establishment treatments when these fertility issues have been remedied.

New Zealand farmers are experiencing a time where there is scrutiny from the general public around farming practices and the environmental degradation, of which they are the perceived cause. The transition of focusing measurements from crop production to soil health and quality shows that the first question has somewhat been answered; now the most pressing issue is maintaining the right to farm. This experiment had become increasingly valuable to collect measurements around sustainability and long-term soil health. Differences in soil quality metrics have emerged between establishment treatments which have functional implications for the management of these treatments. These are primarily an implication of where soil C has accumulated.

The New Zealand Government has proposed the entry of agricultural emissions into the country's Emissions Trading Scheme as an essential step to meet emissions targets of a 30% reduction below 2005 levels by 2030. The relationship between soil C and crop production and particularly the effects of cultivation are not well understood. This experiment may be extremely valuable in understanding the effect of cultivation on agricultural greenhouse gas emissions and quantifying benefits of shifts in practice.

These results are only relevant if the experiment represents actual farming scenarios. Soil fertility has too long been neglected, and it needs to be realigned to what is generally accepted as optimum for the cropping situation. The CET is an extremely valuable resource and has proven useful for measurements that had not been considered at the initiation of the experiment, but over time has provided a unique source of data for soil health under Canterbury arable production.

Care must be taken when adjusting the experiment that the history of treatment differences is not jeopardized and there are no long-standing effects. However, a call must be made when the relevance is being lost, or the experiment can become more powerful through adapting to current circumstances.

References

Barthes, B., Roose, E., 2002. Aggregate stability as an indicator of soil susceptibility to runoff and erosion; validation at several levels. Catena 47 (2), 133–149.

Clifford, P.T.P., 1985. Effect of leaf area on white clover seed production. Grassland Research and Practice Series 2 (6), 25–31.

Cox, J.E., 1978. Soils and Agriculture of Part Paparua County, Canterbury. Wellington: New Zealand Soil Bureau bulletin No. 34. D.S.I.R.

Dersch, G., Böhm, K., 2001. Effects of agronomic practices on the soil carbon storage potential in arable farming in Austria. Nutrient Cycling in Agroecosystems 60 (1–3), 49–55.

Hampton, J.G., Kemp, P.D., White, J.G.H., 1999. Pasture establishment. In: White, J.G.H., Hodgson, J. (Eds.), New Zealand Pasture and Crop Science. Oxford University Press, Melbourne, pp. 101–115.

McKenzie, B.A., Hampton, J.G., White, J.G.H., Harrington, K.C., 1999. Annual crop production principles. In: White, J.G.H., Hodgson, J. (Eds.), New Zealand Pasture and Crop Science. Oxford University Press, Melbourne, pp. 199–212.

McLaren, R.G., Cameron, K.C., 1996. Soil Science: Sustainable Production and Environmental Protection, second ed. Oxford University Press, Melbourne.

Müller, K., Dal Ferro, N., Katuwal, S., Tregurtha, C., Zanini, F., Carmignato, S., De Jonge, L.W., Moldrup, P., Morari, F., 2018. Effect of long-term irrigation and tillage practices on X-ray CT and gas transport derived pore-network characteristics. Soil Research 57 (6), 657–669.

NZ Herald, 2017. Ashburton Farmers Break Guinness World Record for Wheat. Available from: www.nzherald.co.nz/the-country/news/article.cfm?c_id=16&objectid=11830849. (Accessed 22 October 2019).

Poole, N., 2017. New Zealand cropping sequences – where have we come from, what's important for the future?. In: Presentation to Foundation for Arable Research's Arable Research in Action Field Day, Chertsey, 6 December.

Stuff, 2015. World Barley Yield Record Set by Timaru Farmers. Available from: www.stuff.co.nz/business/farming/cropping/68146316/world-barley-yield-record-set-by-timaru-farmers. (Accessed 22 October 2019).

White, J.G.H., 1999. The farmlands of New Zealand. New Zealand pasture and crop science. In: White, J.G.H., Hodgson, J. (Eds.), New Zealand Pasture and Crop Science. Oxford University Press, Melbourne, pp. 1–10.

CHAPTER

Integration of efficient farm enterprises for livelihood security of small farmers

5

S.S. Walia, Tamanpreet Kaur
School of Organic Farming, Punjab Agricultural University, Ludhiana, Punjab, India

Introduction

The Indian population of 1.326 billion is increasing at a rate of 1.6% per annum and is expected to reach 1.705 billion by 2050 (UN, 2015). In addition to this, desertification (6 million ha year^{-1}), soil degradation (5–10 million ha year^{-1}) coupled with declining irrigated area per person (1.3% year^{-1}), and forested area (0.78% year^{-1}) in India are the challenges which need to be addressed together in a system perspective (Panwar et al., 2018). About 65% of the Indian population is dependent on agriculture for livelihood and employment generation. Besides this, there are 129 million farm holdings in the country, out of which 86% are small and marginal (<2 ha). With increasing population, per capita availability of crop land is also decreasing gradually from 0.35 ha in 1961 to 0.12 ha in 2015, which is projected to 0.09 ha per capita in 2050 with the expected country population of 1707×10^6 (Lal, 2016). As the Indian economy is primarily rural and agrarian, the declining trend in size of land holding poses a serious challenge to the profitability and sustainability of farming systems. The crop and cropping system–based perspective of research needs to make way for farming systems–based research conducted in a holistic manner for the sound management of resources available to the small farmers (Jha, 2003). The income and food requirement of the farmers can be augmented and supplemented by the adoption of efficient enterprises such as animal husbandry, horticulture, and fishery along with cropping. An integrated farming system (IFS), therefore, assumes greater importance for sound management of farm resources to enhance the farm productivity and profitability besides reducing environmental degradation. The IFS is practiced in many different countries in many different ways. Yet, a common characteristic of the IFS is a combination of crop and livestock enterprises. Bahire et al. (2010) defined the IFS as an integrated mixed farming system which comprises the practice of raising different yet dependent enterprises and when different enterprises are interdependent they are primarily supplementary and complementary to each other. The IFS takes account of innovation in farming for maximizing production through optimal use of local resources, effective recycling of farm waste for productive purposes, community-led local systems for water conservation,

Long-Term Farming Systems Research. https://doi.org/10.1016/B978-0-12-818186-7.00006-0
Copyright © 2020 Elsevier Inc. All rights reserved.

organic farming, and developing a judicious mix of income-generating activities such as dairy, poultry, fishery, goat rearing, vermicomposting, and others. The IFS approach is considered to be the most powerful tool to enhance the profitability of farming systems, especially for small and marginal farm holders. This chapter deals with dairy-based enterprise combinations for their contribution to sustainable livelihood of farm families with income enhancement as a major plank.

Materials and methods

The present study was conducted at the research farm of the Department of Agronomy, Punjab Agricultural University, Ludhiana, under "All India Coordinated Research Project on Integrated Farming Systems (ICAR)" during 2016–17. An IFS model experiment was initiated during Kharif (March) 2010. The experimental area is located in Central Plain Zone-III of the Punjab State comprising of Ludhiana and situated between the latitude of 30o to 32o56′ N and latitude of 75o to 76o52′ E, with the elevation of 247 m above mean sea level (ICAR, Central Institute of Dryland Agriculture). The soils of the area have developed from alluvium to pleistocons to recent time under ustic moisture regimes (ICAR, Central Institute of Dryland Agriculture). These soils are deep, well-drained, and free from the hazards of salinity and alkalinity. The experimental area is flat upland. The water table is 15 m below the ground level, and the water is free from salts. Both canal and tube well water were used for applying irrigation. The study was conducted on a 1.0 ha model (10,000 sq m area) farm comprising of crops–horticulture–aquaculture–dairy–agroforestry components. In Kharif season, crops grown in the 6400 sq m area were paddy (*Oryza sativa*), maize (*Zea mays*), and turmeric (*Curcuma longa*) and in the following rabi and summer season, potato, fodder berseem (*Trifolium alexandrinum*), wheat, gobhi sarson, onion, pearl millet, and spring maize were grown. The various crop rotations are given in Table 5.1. Around 1600 sq m area was utilized for horticulture practices comprising guava and citrus plantation and the interrow spacing of 1500 sq m area was utilized for raising vegetable crops while 200 sq m, 1000 sq m area, and 300 sq m were meant for dairy, aquaculture, and agroforestry, respectively. In addition to this, boundary plantations with cranberry (karonda) and galgal were also done. The layout of the IFS model is presented in Fig. 5.1.

This manuscript compares the economic viability of the IFS to the prevalent rice–wheat cropping system during 2016–17. The economic analysis of different farming systems including dairy was carried out for which suitable statistical analysis such as percentages and benefit–cost ratio (B:C ratio) was used to establish the proper results and for the proper inference of the study. For comparison between various enterprises, the yields of all enterprises were converted into rice equivalent yield on price basis. For calculating rice equivalent yield, the gross returns of each enterprise were divided with the prevailing market price of rice.

Table 5.1 Crop rotations practiced in integrated farming system.

Cropping system	Kharif	Rabi	Summer
	Maize	Wheat	Summer moongbean
	Maize (cobs) and fodder	Berseem	Baby corn
	Rice	Potato	Onion
	Maize	Berseem fodder and seed	
	Maize + cowpeas	Wheat	Summer moongbean
	Maize	Gobhi sarson	Bajra fodder
	Turmeric	Onion	
	Rice	Potato	Spring maize
(Agroforestry) Litter of poplar	Turmeric	Wheat	
Vegetables as intercrop in horticulture unit	Peas + marigold	Chilli	
	Spinach	Bottle gourd	Bottle gourd
	Cauliflower	Bitter gourd	
	Ladyfinger	Ladyfinger	Cauliflower
	Turmeric	Radish	Radish

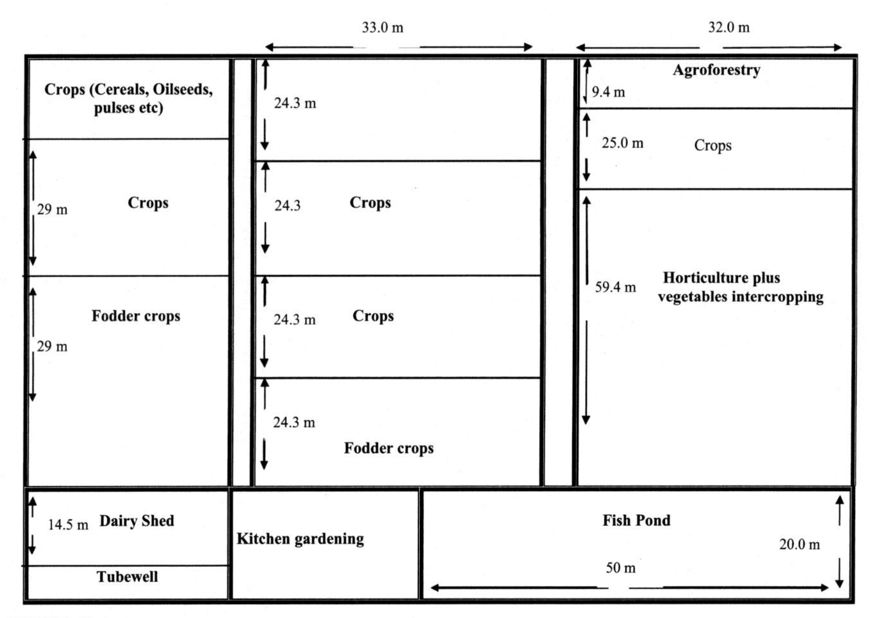

FIGURE 5.1

Layout of integrated farming system model.

Financial gains of adopting different enterprise combinations

In adoption of improved agricultural practices for increasing income of farm families, the farmers are sensitive to the financial gains of the practices. The data from the study indicated that adoption of IFS by inclusion of crop-based enterprises, horticulture, dairy, and aquaculture has recorded overall average net returns of Rs. 401,614 ($5737.34) ha^{-1} with the highest been contributed by dairy (Rs. 186,110 ($2658.71) ha^{-1}) followed by field crops such as maize (*Zea mays*), wheat (*Triticum aestivum*), rice (*Oryza sativa*), summer moongbean (*Vigna radiata*), etc., (Rs. 132,348 ($1890.69) ha^{-1}), horticulture (Rs. 44,860 ($640.86) ha^{-1}), aquaculture (Rs. 25,189 ($359.84) ha^{-1}), and kitchen gardening (Rs. 5600 ($80.00) ha^{-1}), boundary plantation (Rs. 4945 ($70.64) ha^{-1}), and agroforestry (Rs. 2562 ($36.60) ha^{-1}) (Table 5.2). Hence, the 1.0 ha model developed for marginal and small farmers gave gross returns of Rs. 702,341 ($10033.44) ha^{-1}, variable costs (fertilizers, pesticides, feed, seeds, etc.) incurred were Rs. 300,727 ($4296.10) ha^{-1}, and net returns by deducting all variable costs were Rs. 401,614 ($5737.34) ha^{-1}, which were far higher than the prevailing rice—wheat cropping system having net returns of Rs. 109,338 ($1561.97) ha^{-1}. Similarly, component-wise overall average farm production was 46,512.6 REY kg ha^{-1}. Dairy (26,042.8 REY kg ha^{-1}) alone is contributing highest production, followed by crop component (13,161.7 REY kg ha^{-1}), horticulture (4160.2 REY kg ha^{-1}), aquaculture (1866.8 REY kg ha^{-1}), and kitchen gardening (605.9 REY kg ha^{-1}) (Table 5.2 and Fig. 5.2). Apart from growing crop component alone, other subsidiary enterprises are significantly contributing to the net profit of farmer. IFS showed B:C ratio for estimated at 1.34, out of which aquaculture (8.40) component showed the highest B:C ratio followed by horticulture (2.50) and field crops (1.99).

Economic analysis of cropping systems (6400 sq m)

The data presented in Table 5.3 depicted that turmeric—onion cropping system gave maximum net returns of Rs. 26,286 per 0.08 ha followed by rice—potato—onion (Rs. 24,737 ($353.38) per 0.08 ha), maize (cobs), and fodder—berseem—baby corn (Rs. 18,105 ($258.64) per 0.08 ha), rice—potato—spring maize (Rs. 16,772 ($239.60) per 0.08 ha), and maize—berseem fodder + seed (Rs. 14,031 ($200.44) per 0.08 ha). The net returns of other maize-based cropping systems varied from Rs. 10,104 ($144.34) to Rs. 7817 ($111.67) per 0.08 ha. The maize (cobs) and fodder—berseem—baby corn cropping system showed the highest B:C ratio of 3.48 followed by maize—berseem (fodder + seed) cropping system (3.70). The rice—potato—onion showed the highest input costs measured at Rs. 14,536 followed by rice—potato—spring maize (Rs. 12,532 ($179.03) per 0.08 ha), turmeric—onion (Rs. 9560 ($136.57) per 0.08 ha), maize + cowpeas—wheat—summer moongbean (Rs. 7822 ($111.74) per 0.08 ha), maize—wheat—summer moongbean (Rs. 7322

Table 5.2 Financial gains from integrated farming system model.

Farm enterprises	Size of the unit (area/number)	Gross returns (Rs./year)	Cost of production (Rs./year)	Net returns (Rs./year)	B:C ratio	REY (kg ha^{-1})
Field crops (cereals/pulses/oilseeds/green fodders, etc.)	6400 m^2	198,743	66,395	132,348	1.99	13,161.7
Horticulture (guava, lemon, vegetable intercrops)	1900 m^2	62,820	17,960	44,860	2.50	4160.2
Agroforestry	300 m^2	5248	2686	2562	0.95	347.5
Dairy	200 m^2	393,246	207,136	186,110	0.90	26,042.8
Aquaculture (freshwater fish production)	1000 m^2	28,189	3000	25,189	8.40	1866.8
Boundary plantation	–	Galgalkkaronda	0	4945	–	–
Kitchen gardening	200 m^2	9150	3550	5600	1.58	605.9
Integrated farming system model—total allocated land (m^2)	10000 sq m	702,341	300,727	401,614	1.34	46,512.6

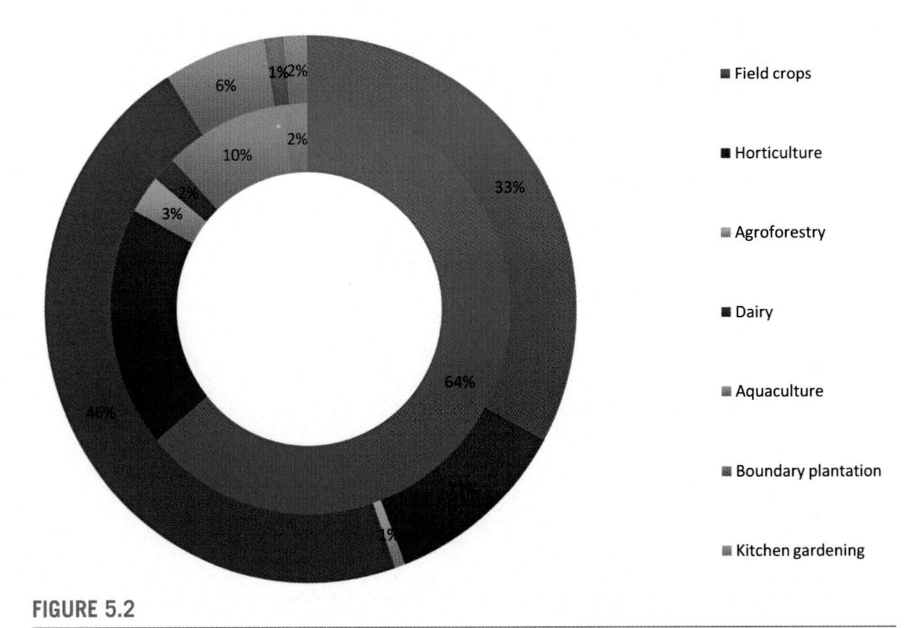

FIGURE 5.2

% Area and their % net returns from different farm enterprises in integrated farming system.

($104.60) per 0.08 ha), maize—gobhi sarson—bajra fodder (Rs. 5634 ($80.48) per 0.08 ha), maize (cobs) and fodder—berseem—baby corn (Rs. 5198 ($74.25) per 0.08 ha), and maize—berseem fodder and seed (Rs. 3791 ($54.16) per 0.08 ha). It was also obtained that rice—potato—onion has generated the highest working days measured at 206 man days followed by rice—potato—spring maize (26 man days), turmeric—onion (16 man days), and maize + cowpeas—wheat—summer moongbean (13 man days).

Economic analysis of horticulture unit (1600 sq m)

The guava and citrus plantation was completed in October 2010 and the interrow spacing of 1500 sq m area was utilized for raising vegetable crops. The data in Table 5.4 revealed that peas + marigold—chilli cropping system gave the highest net returns of Rs. 9698 ($138.54) per 0.03 ha followed by turmeric—radish (Rs. 8613 ($123.04) per 0.03 ha), ladyfinger—ladyfinger—cauliflower (Rs. 6904 ($98.63) per 0.03 ha), spinach—bottle gourd (Rs. 4136 ($59.08) per 0.03 ha), and cauliflower—bitter gourd (Rs. 1409 ($20.13) per 0.03 ha). Peas + marigold—chilli showed not only the highest net returns but also a labor-intensive vegetable-based cropping system by utilizing maximum of 94 man hours. The net returns obtained from all vegetables were Rs. 30,760 ($439.42). The vegetable crops being labor-intensive required 27 man days.

Table 5.3 Economic analysis of different cropping systems followed in integrated farming system model, 2016–17.

Cropping system	System duration (days)	Economic yield kg/800 m^2	Gross return (Rs./0.08 ha)	Input cost (Rs./0.08 ha)	Net return (Rs./0.08 ha)	Total man days	B:C ratio
Maize–wheat–summer moongbean	310	372–495–66	16,570	7322	9248	13	1.26
Maize (cobs) and fodder –berseem–baby corn	170	1975 (794 cobs) –6445–1850 (318)	23,303	5198	18,105	10	3.48
Rice–potato–onion	308	530–1920–1792	39,273	14,536	24,737	26	1.70
Maize–berseem fodder –seed	147	2028–6980–26	17,822	3791	14,031	7	3.70
Maize + cowpeas –wheat–summer moongbean	273	359 + 378–388–74	15,639	7822	7817	13	1.00
Maize–gobhi sarson –bajra fodder	277	380–162–3038	15,738	5634	10,104	11	1.79
Turmeric–onion	335	1722–1872	35,846	9560	26,286	16	2.75
Rice–potato–spring maize	303	534–1880–592	29,304	12,532	16,772	21	1.34
Turmeric–wheat (agroforestry) litter of poplar	358	294–142	5248	2686	2562	6	0.95

Note: Rate (Rs/kg): Rice, 5.10; maize, 13.65; wheat, 16.25; potato, 70; onion, 60; gobhi sarson, 37; fodder (maize, bajra, cowpeas), 1.50; berseem, 1.00; turmeric, 10; baby corn and maize cobs, 10; summer moongbean, 52.25.

Table 5.4 Vegetables in the horticulture unit as intercrop, 2016–17.

Cropping system	Economic yield kg/ 300 m²	System gross return (Rs./ 0.03 ha)	System cost (Rs./ 0.03 ha)	System net return (Rs./ 0.03 ha)	Total man hours
Peas + marigold – chilli	618–48 –512	15,962	6264	9698	12
Spinach–bottle gourd	522–580 –412	6468	2332	4136	5
Cauliflower –bitter gourd	404–318	5041	3632	1409	3
Ladyfinger –ladyfinger –cauliflower	272–280 –460	9890	2986	6904	6
Turmeric –radish	608–478 –518	11,359	2746	8613	1
Total		48,720	17,960	30,760	27

Note: The seasonal vegetables were grown as inter row in horticultural plants.
Rate (Rs/kg): Peas, 12.50; chilli, 11.40; bottle gourd, 5.00; cauliflower, 5.00; bitter gourd, 9.50; lady-finger, 10.00; radish, 5.30.

Economic analysis of dairy component (200 sq m)

There are two cows in the dairy component of IFS. The breed of one cow is Holstein Friesian (HF) and other is Sahiwal. The milk yield varied from 550 to 1209 L/month. The total milk yield was 12,635 L during 2016–17. The data presented in Table 5.5 depicted that net profit earned from dairy component after deducting the input costs (Rs. 207,136; $2959) amounted to Rs. 186,110 ($2659) and the gross income earned was Rs. 393,246 ($5618). B:C ratio for dairy component was 0.89. Apart from the milk yield data, from our experience with cows of two different breeds we conclude that the HF cow is very delicate and requires extra care and therefore not suited for small and marginal farms in Punjab region. Furthermore, the data presented here were obtained. If the farmer manages the dairy unit with family labor, then profit can be further increased.

Economic analysis of aquaculture (1000 sq m)

Rohu, catla, and mrigal fingerlings were released during December 2015 and 220 kg fish was harvested during November 2016. The washings from dairy shed are managed to put into fish pond which serve as a source of nutrients for fish. Table 5.2 showed that the aquaculture unit after deducting the cost of production (Rs. 3000 year^{-1}; $42.85) fetched gross returns of Rs. 25,189 ($359.84) with B:C ratio of 8.40.

Table 5.5 Economic analysis of dairy component in integrated farming system, 2016–17.

Particulars	Amount (Rs.)	Amount ($)
Income generated by sale of milk (9381 L @ Rs. 40/L)	375,246	5361
Potential income that can be generated through sale of one HF Heifer, one sahiwal Heifer, and two HF female calves (value not included in economics)	130,000	1857
Cow dung used to prepare FYM	18,000	257
Gross income	393,246	5618
Cost of feed (Rs.)	146,000	2086
By-pass fat (fat-enriched ingredient) and minerals	19,564	279
Medicine and other misc charges	12,000	171
Labor cost @ 2 working hours day^{-1} (Rs. 35/hr^{-1})	29,572	422
Input cost (Rs.)	207,136	2959
Net profit	186,110	2659
B:C ratio	0.89	

Economic analysis of agroforestry (600 sq m)

Poplar (*Populus nigra*) plants have been planted during the month of February 2011 and turmeric–wheat was grown as interrow crop which gave turmeric and wheat yield of 294 and 142 kg. The gross return from turmeric was Rs. 2940 ($42.00) and from following wheat was Rs. 2308 ($32.97) and cost incurred was Rs. 2686 ($38.37) and net returns were Rs. 2562 ($36.6). Now the value of 21 poplar plants is approximately Rs. 42,000 ($600.0).

Recycling of residues in integrated farming system

The straw was used to feed to the cows, heifer, and two young calves. The residue of other crops viz., rice were utilized for bedding in the cattle shed during winter months and then these rice residues were decomposed in the vermicompost pit along with cow dung. The remaining residues of various crops (3913 kg) were utilized for mulching. On an average, these residues contained 0.6%, 0.4%, and 1.2% of N, P, and K, respectively, which amounts to 23.5, 15.6, and 47 kg of N, P, and K, respectively. These residues contained N, P, and K amounting to Rs. 295 ($4.21), 697 ($9.95), and 862 ($12.31). In a nutshell, these residues contained fertilizer amounting Rs. 1854 ($26.48), which upon recycling will improve soil health as well (Table 5.6).

Table 5.6 Straw/stover produced (kg/800 sq m area) and utilized in integrated farming system (IFS).

Cropping system	Straw/stover (kg/800 sq m area)				
	Kharif	Rabi	Summer	Used as input in IFS	Used as nutrient source
Maize–wheat–summer moongbean	668	782	226	1,450	226
Maize–berseem–baby corn	0	0	0	0	0
Rice–potato–onion	736	524	0	736	984
Maize–barseem fodder and seed	709	0	0	709	0
Maize + cowpeas –wheat–summer moongbean	0	732	248	732	248
Maize–gobhi sarson –bajra fodder	725	517	0	725	517
Turmeric–onion	688	0	0	0	688
Rice–potato–spring maize	788	462	840	840	1,250
Turmeric–wheat (agroforestry) Litter of poplar	118	178	0	178	118
FYM					18,000

Conclusion

Rice–wheat cropping system obtained net returns of Rs. 109,338 ha^{-1} ($1561.97) (Table 5.8). Moreover, the returns are obtained two times annually. However, in IFS approach, a farmer can obtain net returns of Rs. 401,614 ha^{-1} ($5737.34), which is distributed throughout the year. Moreover, the dairy unit provides income on daily basis. Vegetable and fruit component provides balanced nutrition to the family members. In the rice–wheat system, Rs. 2300/ha day net returns can be obtained, whereas from IFS the net returns were Rs. 1100/ha day, i.e., Rs. 800/ha day extra over rice–wheat cropping system. The net returns obtained from IFS model since its inception are also presented in Table 5.7 and they showed an increasing trend over time as shown in Fig. 5.3, indicating the viability and profitability of IFS. Thus, it is essential that IFS should be supported through institutional, extension, policy, and marketing interventions so that the adoption of multiple farm enterprises in an integrated manner can ensure a substantial income generation to sustain the livelihood of farmers over the meager income from self-standing enterprises as revealed from this study.

Table 5.7 Net returns of integrated farming system since inception.

Year	Net returns (Rs/ha)	Net returns ($)
2016–17	401,614	5,737
2015–16	400,870	5,727
2014–15	380,308	5,433
2013–14	328,191	4,688
2012–13	202,551	2,894
2011–12	155,207	2,217
Average (6 years)	311,457	4,449

Table 5.8 Grain yield and economic analysis of rice–wheat cropping systems.

Cropping systems	Rice grain yield (kg ha^{-1})	Wheat grain yield (kg ha^{-1})	Gross returns (Rs./ha)	Input cost (Rs./ha)	Net return (Rs./ha)
Rice –wheat	6,555	5,162	182,863	73,525	109,338

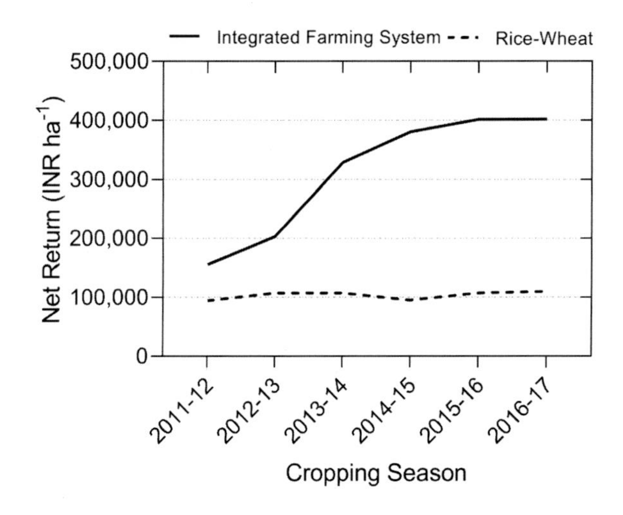

FIGURE 5.3

Line graph showing net returns of integrated farming system model since inception and rice–wheat cropping system.

References

Bahire, V.V., Kadam, R.P., Sidam, V.N., 2010. Sustainable Integrated Farming is the need of the Indian farmer. In: 22nd National Seminar on Role of Extension in Integrated Farming Systems for Sustainable Rural Livelihood, 9th–10th Dec, Maharashtra, p. 65.

Jha, D., 2003. An overview of farming systems research in India. Annals of Agricultural Research 24 (4), 695–706.

Lal, R., 2016. Managing soil and water resources for sustainable intensification of agro-ecosystems in India. Indian Journal of Fertilizers 12 (11), 18–29.

Panwar, A.S., Shamim, M., Ravisankar, N., 2018. Opportunities and challenges of doubling farmers' income in Indo-Gangetic Plains through integrated farming systems. In: XXI Biennial National Symposium of Indian Society of Agronomy, 24–26 October, 2018. MPUAT, Udaipur, Rajasthan.

SECTION

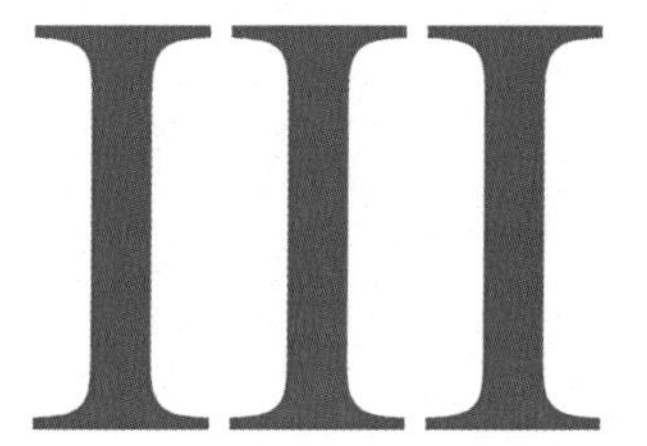

Maximizing outcomes through networking and capacity development

CHAPTER

Building capacity from Glenlea, Canada's oldest organic rotation study

6

Michelle Carkner, Keith Bamford, Joanne Thiessen Martens, Sarah Wilcott, April Stainsby, Katherine Stanley, Calvin Dick, Martin H. Entz

Department of Plant Science, University of Manitoba, Winnipeg, MB, Canada

Introduction

In Canada, crop rotation studies began in the early 1900s; researchers saw the value in creating experiments to learn about crop production and how agricultural practices impact the soil (Poyser et al., 1957; Campbell et al., 1990; Janzen, 1995). The growth of the organic agriculture industry justified a dedicated organic long-term study, which led to the establishment of the Glenlea long-term crop rotation study (referred to as the Glenlea study) in 1992 and is Canada's oldest organic field crop study. The Glenlea study is located 20 km south of Winnipeg, Manitoba, on Treaty 1 territory. The soil is a Rego Black Chernozem comprised of 12% sand, 32% silt, and 55% clay, and it has an organic matter content of 5.5%.

Agronomists, crop ecologists, and entomologists were involved in planning the Glenlea study. They were concerned with the sustainability of fertilizer and pesticide use and wanted to determine if proper crop rotation could reduce the need for inputs or eliminate them entirely (Entz et al., 2014). The original experimental design was a split plot randomized complete block with three replicates. The main plots had three 4-year crop rotation treatments that included (i) grain-only (wheat−pea−wheat−flax); (ii) green manure−grain (wheat−clover green manure−wheat−flax); and (iii) forage−grain (wheat−alfalfa−alfalfa−flax). Flax has a low degree of weed competitiveness; therefore, it was grown in all rotations as a "test" crop. Subplots consisted of four crop input combinations and each block contained a restored native prairie grass as an ecological benchmark. The crop input treatments were: (i) fertilizer and herbicide added (+f +h) (later termed "conventional"); (ii) only fertilizer added (+f−h); (iii) only herbicide added (−f +h); and (iv) no inputs (−f−h) (later termed "organic").

The Glenlea study was modified after harvest in 2003, in preparation for 2004 planting, to the current experimental design. The two main rotation types are "grain-only" (flax−oat−soybean−wheat) and "grain-forage" (flax−alfalfa−alfalfa−wheat). The plots were split to allow the rotation to be fully phased, meaning that all crops in the rotation are grown each year. Both rotation types are grown under conventional and organic management. A green manure plow down is substituted for soybean in

Long-Term Farming Systems Research. https://doi.org/10.1016/B978-0-12-818186-7.00007-2
Copyright © 2020 Elsevier Inc. All rights reserved.

the organic grain-only rotation. The current design still includes a native grass prairie in each of the three replicates. A depiction of the changes is shown in Fig. 6.1. In short, the larger blocks of +f/+p and −f/−p (deemed the conventional and organic treatments, respectively) within the first two ranges of the study were each divided into 4 by 30 m

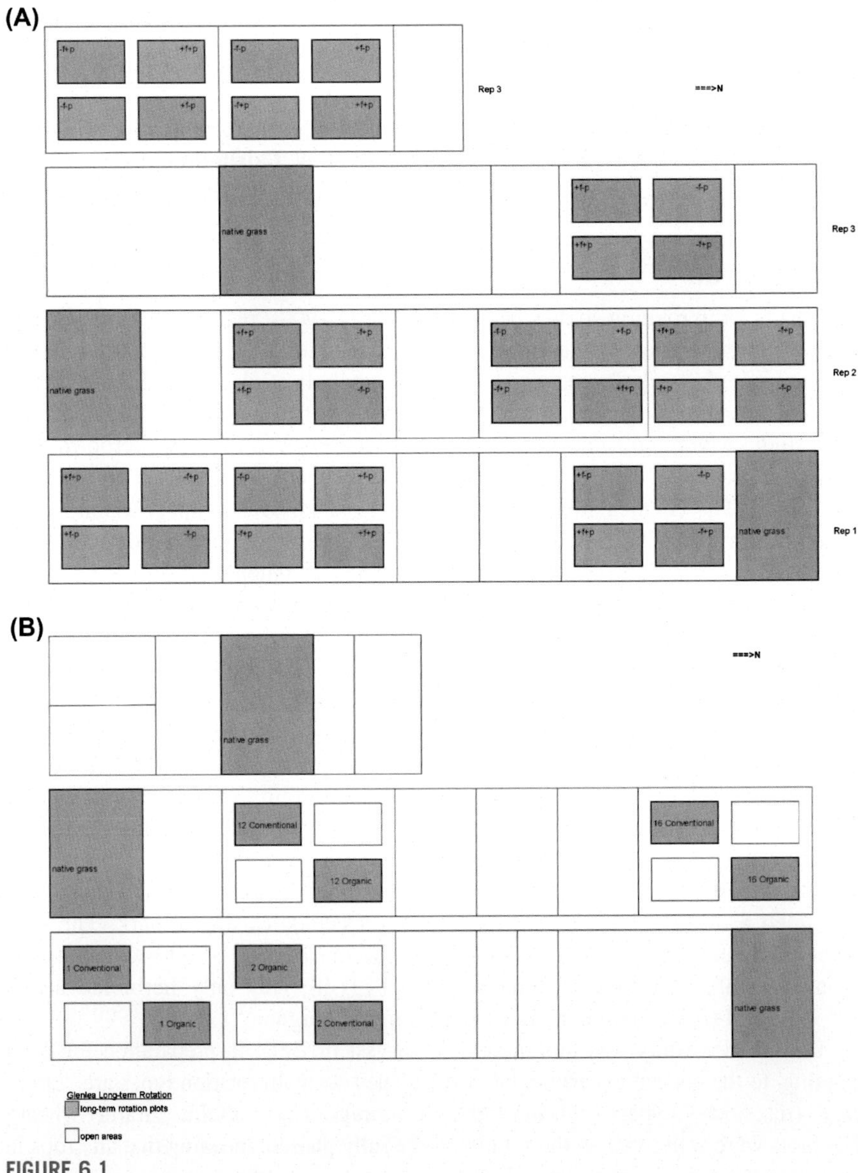

FIGURE 6.1

The original plot layout (A) and the current plot layout (B) for the Glenlea long-term rotation. In the new layout, plots 1 and 16 include grain-only rotations, while plots 2 and 12 include forage–grain rotations.

plots. These plots are separated by a 2 m space between plots, which is regularly tilled to reduce weed encroachment. These 4 by 30 m plots were assigned different phases in the rotation so that (1) all rotation phases appear each year and (2) three replicates of all treatments are distributed across the two ranges resulting in a completely randomized design for each of the two rotations (i.e., forage–grain and grain-only rotations).

Weed control practices also evolved over time. Postemergence harrowing was practiced in early years, but crop damage was deemed excessive in this vertisol soil, so the practice is only used in years when no soil clods appear and rainfall after harrowing is not imminent. Beginning in 2011, interrow cultivation is practiced in all grain crops. When wheat, oats, and flax crops are approximately 15 cm tall, an implement with a backswept knife is used to kill weeds growing between the rows.

The research done at this site has greatly enhanced our knowledge of organic systems in Canada and continues to produce valuable research, training, and learning by scientists, students, agronomy advisors, and farmers. In this chapter, we focus on the major findings from Glenlea, asking questions about how many years were required to identify particular trends. We then describe how these major findings shaped new research and graduate training projects, as well as educational and outreach programs for the Canadian organic sector.

Identification of major limitations to organic production based on the Glenlea study

Phosphorus depletion in the hay export rotation

Signs of nutrient deficiencies in the forage–grain organic rotation were apparent after year eight, as evidenced by declining alfalfa yields and reduced soil available phosphorus (P) levels (data not shown). A significant interaction between rotation and inputs in year 12 (Table 6.1) showed that P was significantly lower in the organic forage–grain system compared with the conventional, but no differences were observed between the two other crop rotations. This was further evidence that the hay export system was depleting available soil P.

The P depletion observation raised questions about the future sustainability of hay export from the forage–grain organic system and confirmed findings from previous studies and surveys in the region. In one example, a rotation study that began in 1958 at Indian Head, Saskatchewan, showed that wheat yield declined in a forage–grain system, where P replacement was withheld, and this effect was quite strong after 30 years (Campbell et al., 1990). A survey of commercial organic farms in the late 1990s showed low levels of available P, especially when they did not include manure additions (Entz et al., 2001); this was confirmed in a follow-up study 10 years later (Knight et al., 2010). The observations of apparent P decline at Glenlea, together with knowledge gained from previous research and surveys, prompted a detailed study in the 13th year of the study, where the extent and nature of the P limitation in the hay export organic system was examined in detail (Welsh et al., 2009).

Table 6.1 Soil nutrient status (kg ha^{-1}) for the Glenlea long-term cropping systems study flax test crop in 2003.

Rotation	Inputs	N[a]	P[b]
Annual	Conventional	32	46
	Organic	22	33
Green manure	Conventional	29	24
	Organic	31	37
Forage	Conventional	81	24
	Organic	37	11
Rotation (R)		0.0024	0.0020
Inputs (I)		0.0093	0.1899
R x I		0.0158	0.0153

[a] *Sampling depth 0–60 cm.*
[b] *Sampling depth 0–15 cm.*

In response to low P, the decision was made to supplement the organic plots with manure additions. All organic plots were split into manured and unmanured treatments in 2007. Composted beef cattle manure (20 Mg wet weight ha^{-1}) was added to wheat stubble in the annual organic grain system in 2007 and on alfalfa in the organic forage–grain system once per rotation cycle beginning in 2007. Typical characteristics of the compost were N = 25.2, $P = 5.0$, K = 24.5, S = 2.5, and dry matter = 490 g kg^{-1}. The use of manure to replenish fertility is widely used in long-term field studies such as the Rothamsted experiment established in 1843 in Britain, the DOK trial established in 1978 in Switzerland, and the Alternative Cropping System trial initiated in 1994 in Scott, Saskatchewan (Mäder et al., 2002; Lemke et al., 2012; Johnston and Pulton, 2018). It was hypothesized that manure addition would revive alfalfa productivity and in turn restore the productivity of the subsequent crops, especially in the hay export system.

Comparing wheat yields in the manured and unmanured forage–grain rotation showed that alleviating the nutrient deficiency through manuring provided an immediate yield correction in the organic production system, producing organic wheat yields that were, on average, only 24% lower than conventional wheat yields (Fig. 6.2). Crop response to manure often takes a number of years because only a percentage of the nutrients are available to plants in the first year (Eghball et al. 2002, 2004). Composted manure has a lower level of nutrients available at the time of application but provides more in subsequent years when compared with fresh manure (Blackshaw et al., 2005). A large percentage of N in manure is in an organic form and therefore not available to plants right away; however, P is mainly inorganic, so most is available to plants in the year of application (Eghball et al., 2002), which may explain the very fast response that was seen in the P-depleted soils (Fig. 6.2).

We attributed wheat yield improvements from manuring to two factors. The first and most important effect was to correct the P deficiency in the alfalfa crop. This had

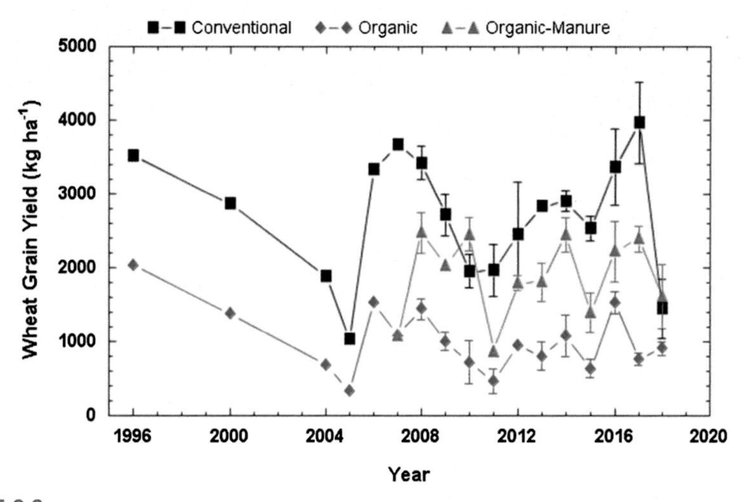

FIGURE 6.2

Wheat yields from the Glenlea long-term forage–grain rotation from 1996 to 2018.

the effect of increased alfalfa yield; the manured plots increased yield on an average of 1694 kg ha^{-1} compared with nonmanured plots for year 1 alfalfa, and 3905 kg ha^{-1} for year 2 alfalfa (Entz et al., 2014). Alfalfa yields from the manured plots became similar to that of the conventional management, and in some years, the manured plots had markedly higher alfalfa yields than conventional management. A depiction of the dramatic effect of manure on alfalfa growth is shown in Fig. 6.3.

FIGURE 6.3

Alfalfa crop growth in manured and unmanured treatments, 2013.

Photo credit M. Entz.

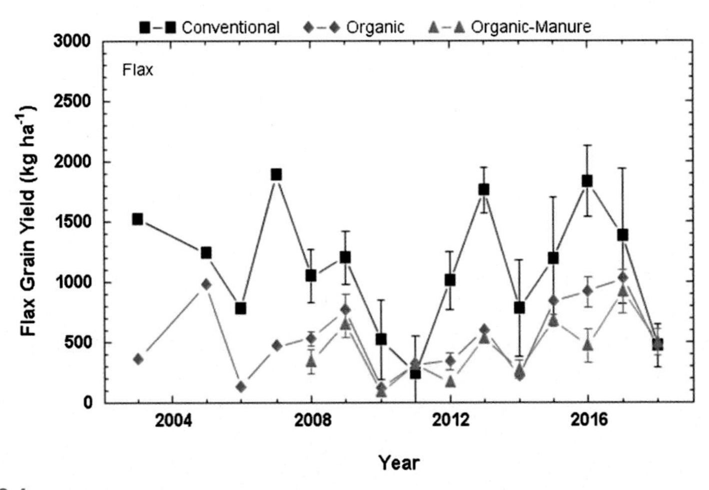

FIGURE 6.4

Flax yields from the Glenlea long-term forage–grain rotation from 2003 to 2018.

Greater alfalfa growth due to manure (Fig. 6.3) appeared to increase N supply to the following crop, which was wheat (Fig. 6.2). We developed the "P increases N supply" theory at Glenlea. That is, correcting the P deficiency in the alfalfa allowed it to increase production, resulting in more N fixation and N available to the following crops.

The second long-term contribution of the manuring was attributed to an N boost to crops. This is evidenced by the immediate improvement in yield of wheat in 2008, the year after autumn manure application (Fig. 6.2).

Greater levels of soil biological activity insufficient to supply P

Soil biological parameters were measured at Glenlea in several different studies (e.g., Braman et al., 2016). It was interesting to observe that compared with the conventional treatments, even the P-depleted (i.e., no manure) organic forage–grain rotation had more favorable levels of arbuscular mycorrhizal fungi (AMF) (Entz et al., 2004; Welsh et al., 2009), phosphatase enzyme activity (Fraser et al., 2015), microbial biomass C (Braman et al., 2016), and bacterial community diversity (Li et al., 2012). Others have also measured improvements in soil health dynamics in organic compared with conventional rotations; for example, Mäder et al. (2002) in the DOK (biodynamic, organic, and conventional) trial in Switzerland. But, despite higher levels of soil health indicators for the organic no manure forage–grain system at Glenlea, crop productivity had virtually collapsed, especially for wheat (Fig. 6.2) and alfalfa (Fig. 6.3). These results clearly show that in certain circumstances, improvements in soil health indicators do not necessarily translate to improvements in crop productivity.

These results have helped farmers and farm advisors better understand the relative importance of nutrient replacement and soil biology indicators in organic crop production.

Manure response absent in organic grain-only rotation

While wheat in the forage–grain rotation responded to added manure, wheat in the grain-only organic rotation did not. For example, between 2008 and 2016, there were no significant wheat yield differences between the manured and control plots for the grain-only rotation. Furthermore, wheat yields in the grain-only organic system were similar or greater than in the manured forage–grain system (2029 kg ha^{-1} and 1950 kg ha^{-1}, respectively). This suggests that nutrients such as P did not appear limiting wheat in the grain-only organic system even after 25 years.

Two factors should be considered in interpreting these results. First, manure was only added to the grain-only organic rotation once, in 2007. Second, at the outset, soils at Glenlea were very high in P and none of the P pools in the grain-only rotation investigated after year 13 showed any signs of depletion (Welsh et al., 2009).

Trade-off between nutrient correction and weeds

While the addition of manure to alfalfa increased organic wheat yield, flax yields were not improved with manure (Fig. 6.5). In fact, in some years, manure addition to alfalfa resulted in significant flax yield reductions, despite the fact that interrow weeds were partially controlled with in-crop tillage. Two explanations were developed for these observations. First, manuring increased weed biomass in the flax crops (Fig. 6.6).

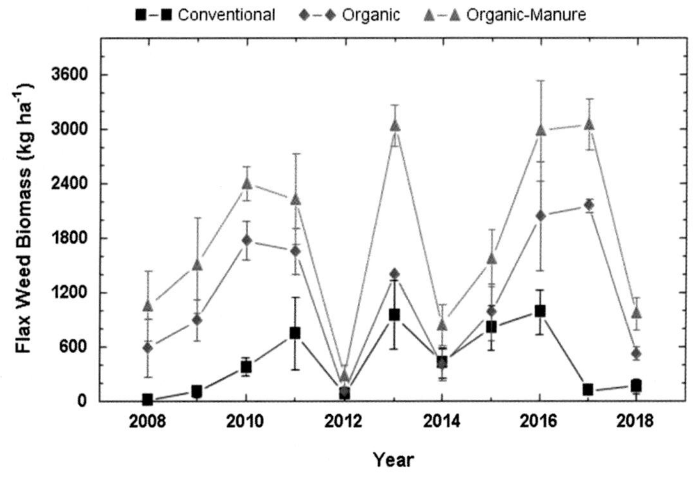

FIGURE 6.5

Weed biomass in flax phase of the forage–grain rotation in the Glenlea study.

FIGURE 6.6

Unmanured (back) and manured (front) flax plots in the forage–grain rotation in 2015.

Photo credit M. Entz.

Therefore, weeds responded better than flax to the added nutrients. Gruenhagen and Nalewaja (1969) observed a better response of wild buckwheat to nutrients compared with flax. The second explanation regards AMF. Flax is highly mycorrhizal, while the dominant weed at Glenlea (wild mustard) does not readily make mycorrhizal associations (Jordan et al., 2000). Also, flax relies more readily on mycorrhizal associations for nutrient uptake than wheat (Grant et al., 2009); therefore, we would expect any AMF interactions to be potentially stronger with flax than wheat. Previous assessments of AMF at Glenlea have confirmed that mycorrhizal colonization is greater in the organic flax compared with the conventional flax (Entz et al., 2004). Taken together, these reasons may explain why flax did not require added manure to achieve a yield increase and why weed competition was greater in the manured compared with the unmanured treatment (Fig. 6.6).

A "1 year in four" legume-based green manure insufficient for N supply

Evidence from Glenlea pointed to an apparent lack of N supply in the organic grain-only system. This deficiency was evidenced in a number of ways. The first was from the "intermediate treatments," where combinations of plus/minus fertilizer and herbicide treatments were applied in the first 12 years of the study (Table 6.2). For example, the difference in flax yield in the full input compared with minus fertilizer/plus herbicide treatment in the 4th, 8th, and 12th year of the study

Table 6.2 Grain yield (kg ha^{-1}) for flax "test crops" at Glenlea for 1995, 1999, and 2003.

		Yield (kg ha^{-1})		
Rotation	**Inputs**[a]	**1995**	**1999**	**2003**
Green manure–grain rotation	+f + h	1808.6	1826.7	971.0
	+f–h	1233.7	1099.8	77.0
	–f+h	1109.9	1584.3	670.0
	–f–h[b]	1022.8	993.4	170.0
Forage–grain rotation	+f+h	1712.3	1453.9	1328.0
	+f–h	1291.8	998.3	94.0
	–f+h	1550.4	1531.8	1287.0
	–f–h	1373.3	1378.9	482.0
Rotation (R)		0.0747	0.0915	0.0010
Management (M)		0.0001	0.0001	0.0010
RxM		0.1575	0.1922	0.0010
SEM		117.01	156.12	615.00

Statistical analysis for yield performed on log transformed data. Means considered significant when P < .05.
[a] *± refers to either addition (+) or omission (–) of fertilizer of herbicide.*
[b] *Renamed "organic grain-only."*

was 162, –78, and 41 kg ha^{-1} for the forage–grain rotation compared with 699, 242, and 301 kg ha^{-1} in the grain-only rotation. That is, withholding fertilizer additions, which included N and small amounts of seed-placed P, had a much greater negative effect in the grain-only than in the forage–grain rotation (Table 6.1). Phosphorous was ruled out as a possible source of the problem as no P deficiency was observed.

A second reason for concluding an N deficiency in the grain-only rotation was low grain yield of oat, the crop which appeared in the final year of the 4-year grain-only rotation since 2004. For example, average oat yield for the period 2008–2018 was 1809 kg ha^{-1} in the organic system compared with 3592 kg ha^{-1} in the conventional system.

Taken together, this evidence strongly suggested that a green manure legume "plowdown" crop once every 4 years was insufficient for supplying adequate levels of N under conditions experienced at Glenlea. This observation is highly relevant to commercial organic production in the region because farmers, while using green manure legume phases, often do not include them more than 1 year in 4 or 5 (Nelson et al., 2010). Infrequent use of green manures in rotation may be because of economic constraints, as it requires farmers to lose a year of income from a cash crop, in addition to the cost of seed, seeding, maintenance, and termination (McCartney et al., 2010; Thiessen Martens et al., 2015). Others also have observed that growing sufficient green manure dry matter to support the nutrient requirements of cash crops remains a challenge (Peoples et al., 2001).

Based on this, a complimentary study was established in 2003 to test new green manure systems to overcome N deficiencies in organic grain-based cropping systems. The 5 ha area, called the "Organic Crops Field Laboratory" explored green manure species choice, reduced tillage green manure termination strategies and green manure grazing management.

Building capacity

The Glenlea study has been responsible for building capacity of students, educators, technical staff, farmers, and advisors. Table 6.3 outlines how the major findings at Glenlea were leveraged to building this capacity. The remainder of this chapter will describe these efforts.

Nutrient monitoring tools

The Glenlea study, as well as other long-term crop and soil management studies on the Prairies (e.g., Campbell et al., 1990), provided evidence that low soil P can limit crop production. However, the ability to predict *when* declines would be observed has not been well established, yet such information is of importance to farmers and farm advisors currently managing Canadian organic systems. At the same time, it has been documented that despite low soil-P test results, organic farmers on the Canadian prairies often experience acceptable yields (Martin et al., 2007). This trend was confirmed at Glenlea where wheat yield was affected by low P (Fig. 6.2) while flax yield was not (Fig. 6.4). Many organic farmers attribute soil biological activity to their acceptable yields; and while this may be true for mycorrhizal crops such as flax (Entz et al., 2004), soil biological activity does not explain why productivity declined in the annual phases of the organic forage–grain rotation at Glenlea (Braman et al., 2016). Traditional soil tests do not appear to provide a complete picture of the soil fertility status on organic farms, as nutrients are slowly released through soil biological processes over time. For these reasons, two new nutrient management tools were developed, having been inspired from experiences at Glenlea.

The green manure bioassay

Using plants to detect available soil-P can be more sensitive than the Olsen-P soil test (Huang et al., 2005), the standard soil test for available P on organic farms in the region. However, a plant bioassay had never been explored on Canadian prairie organic farms. Therefore, a knowledge gap presented itself—could we equip organic farmers and advisors with a tool to avoid the declines observed in the Glenlea rotation before it was too late?

To create a benchmark of soil-P status versus plant response, 41 fields on 18 organic farms across the Canadian prairies were sampled between 2014 and 2016 (Thiessen Martens et al., 2019). Plant tissue tests from green manure crops (including weeds) were completed with specific interest in P such that comparisons between soil-P test results could be compared with P plant uptake.

Table 6.3 A summary of limitations discovered at Glenlea and how these resulted in capacity building in research, outreach, and their impact on the wider agricultural community.

Documented limitation	Capacity building in research	Capacity building in practice	Impact
P limitation in organic systems with high P export	Research into manure as source of P for organic farms. 1 PhD; 1 MSc	Green manure bioassay and nutrient management budget in organic agronomic training program	Greater capacity to make informed production system decisions, think critically about current issues in agriculture and beyond. Number of papers published on phosphorus at Glenlea: 7, cited 147.
N limited productivity in the organic grain-only system	Green manure research, grazed green manures, reduced tillage green manure management, green manure economics, and disease in green manure systems. 2 PhD; 3 MSc	Field days, website green manure productivity tool, Organic Agronomist Training. Green manure tool issued by Prairie Organic Grain Initiative (POGI)	Improved green manure species suitability and termination knowledge. Number of papers published on green manure and no-till research: 11, cited 88.
Improved soil biology alone not enough improve production	Research into manure as source of P for organic farms. 3 MSc	Education on interpreting soil biology information for agronomists and farmers to possibly avoid a breakdown in the organic system, in addition to field days, website, etc.	Farmers are able to make informed decisions about cropping systems, purchasing, and decision-making. Papers written on subject 2; cited 41 times.
Trade-off between nutrient supplementation and weed biomass	Research into alternative weed control methods. New infrastructure grant for state-of-the-art equipment	Organic Agronomist Training, useful information for organic farmers, in addition to field days, webinars, etc.	With greater knowledge of weed dynamics, farmers can make informed decisions about management and may be able to prevent problems seen at Glenlea.

Results from the 41 survey fields showed that soils with available P levels above 15 ppm always had a plant tissue P concentration above the minimum threshold (i.e., 0.2% P, Thiessen Martens et al., 2019). Results also showed that when soil test P was below 5 ppm, plant tissue P was always below the 0.2% P threshold (Fig. 6.7). Based on these observations, we concluded that the Olson-P soil test adequately described P status for organic fields either below 5 ppm or above 15 ppm. This information has become useful for crop advisors and farmers, who are now more confident about when the "standard" Olson-P soil test is most able to predict P availability.

But what about soil test P levels between 5 and 15 ppm? It is within this zone where the green manure bioassay appears to have its greatest value. For example, in our survey, almost all of the green manure crops with soil available P between 5 and 10 ppm showed plant tissue P concentrations above the 0.2% P threshold (Fig. 6.7). This observation supports what many farmers and crop advisors have experienced; that despite lower available soil P levels (between 5 and 15 ppm), crops are still able to acquire P and accumulate tissue P above the minimum level. This challenges the dominant narrative among conventional crop advisors—that when soil P is below 15 ppm, farmers will have a problem. Based on our results, it appears that the green manure bioassay tool can be helpful in predicting P sufficiency when soil P levels are "marginal," and in these cases, the green manure bioassay appears more useful than soil testing.

The nutrient budgeting tool

Results from Glenlea inspired the development of a second tool. The nutrient budgeting tool is an excel-based calculator that considers all nutrient imports and exports from individual fields on a farm. Input and output values are based on lab

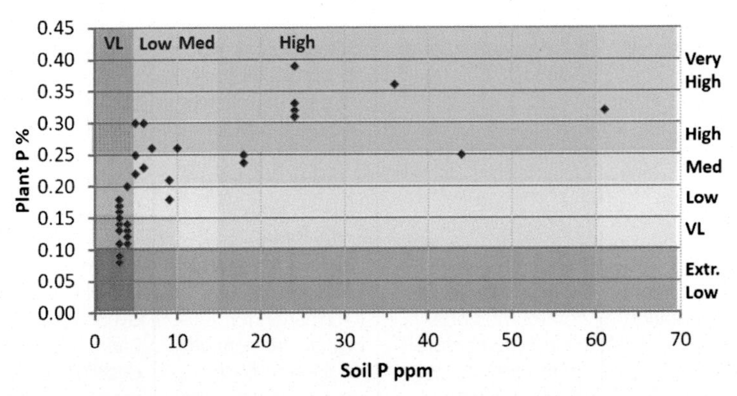

FIGURE 6.7

Relationship between soil P concentration and plant tissue P concentration in green manures grown on Manitoba and Saskatchewan organic farms in 2015 and 2016. Agronomists were encouraged to advise farmers to apply manure at the "low" and "VL (very low)" plant P %.

Image: Joanne Thiessen Martens.

analysis of plant, seed, manure, etc. The nutrient budgeting tool uses "book values" where lab assessments are not possible, and when farmers choose not to pay for the extra analysis costs. N fixation through legumes can be estimated from legume biomass information collected through the green manure bioassay tool. Similar nutrient budgeting tools have been created in Europe (Bachinger et al., 2012); however, to our knowledge, this is first such tool for the Canadian Prairies.

Direct training of farm advisors

To test the efficacy and ease of use of our tools, a pilot project was initiated by the Prairie Organic Grain Initiative (POGI). Fourteen agronomists were partnered with 52 farmers across the Prairies to carry out bioassays on green manures and create a nutrient budget alongside the farmer. Agronomists were encouraged to take a "codesign" approach with the farmers as outlined in Fig. 6.8. The pilot project was part of a larger education program called "Organic Crop Agronomist Training" that serviced the prairies with the goal to train practicing agronomists how to properly and effectively advise organic farmers. Although the curriculum was designed for agronomists, organic farmers registered for the training as well. In total, 161 participants from British Columbia, Alberta, Saskatchewan, Manitoba, and northern Ontario received training through six in-person workshops in Alberta, Saskatchewan, and Manitoba, participated in online discussions via a closed Facebook group, and joined 13 phone-in seminars called "Coffee Shop Talks." Lessons learned from the Glenlea study, the bioassay tool, and the nutrient budgeting tools were used as a major platform in the curriculum.

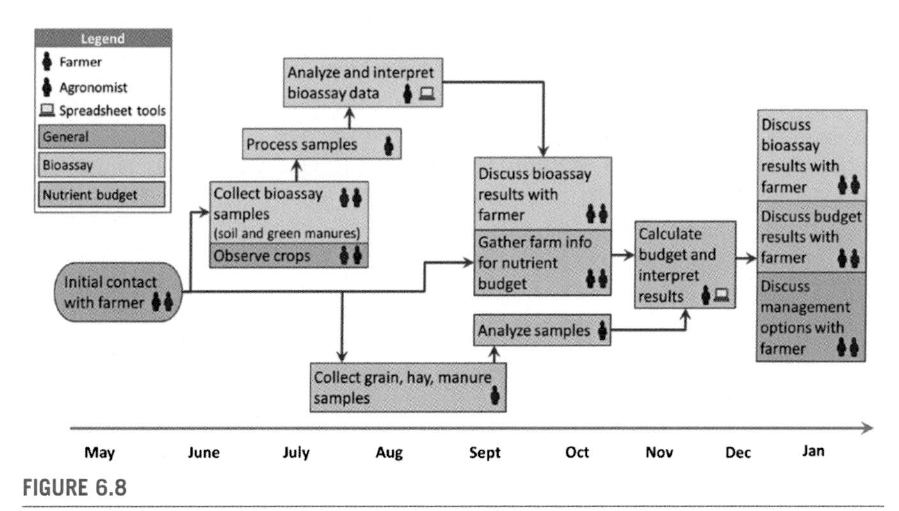

FIGURE 6.8

Outline of the "codesign" approach agronomists in the program were encouraged to take with farmers.

Image: Joanne Thiessen Martens.

Many agronomists reported that the program gave them more confidence when working with organic farmers: "The resources were priceless, and put agronomists in a position to be confident and successful." Others stated that the nutrient budgeting tool caused farmers to change their practices: "The change in cattle management and the decision to import off-farm manure was the biggest change in farm management as a result of the nutrient management tool." One agronomist described that the farmer found that "[The tool was] very valuable. Information and advice that he wouldn't have received if he had not participated in this program."

The information was there, but they had the resources to compile and interpret it. This program has long-term production value/implications for the future.

Building knowledge capacity of green manure management

When the Organic Crops Field Laboratory was established, little was known about the potential use of different legumes as green manures in Manitoba. It had been reported that sweet clover and alfalfa were the most common green manures used on organic farms (Entz et al., 2001; Knight et al., 2010); however, the efforts for nutrient building may be fruitless if the farmer is harvesting for forage. In addition, some farms were reported to have no soil-building crops in their rotation at all (Entz et al., 2001). It is possible that many organic farmers believed that they did not need a soil-building phase in their rotation because of acceptable yields in the short term; however, results from the Glenlea study demonstrated that a productivity decline is inevitable in some cases.

Major criticisms of green manure use in rotation include use of tillage for termination, reduction of available N for the next crop, N leaching, and weed control. To address these concerns, a range of no-till or reduced tillage termination strategies were investigated for organic production. Nitrogen leaching potential, N availability to the following crop, and subsequent crop performance (wheat, dry edible beans), as well as weed dynamics were examined in relation to the termination tools—blade roller, undercutter, flail mower, and a combination of blade rolling and tilling in the fall or spring (Vaisman et al., 2011; Halde and Entz, 2014; Evans et al., 2016; Podolsky et al., 2016). We found that fall tillage could be reduced by rolling the cover crop instead of cultivating it (Podolsky et al., 2016). We established mulch quality and decomposition rates of six different annual cover crop species in organic no-till systems (Halde and Entz, 2014). In addition, we tested 10 different combinations of green manures under organic no-till conditions and reported that pure hairy vetch or a barley/hairy vetch intercrop was the most favorable for Manitoba's growing environment for its ability to produce acceptable wheat yields and suppress weeds (Halde and Entz, 2014). In our experience, hairy vetch's high biomass accumulation potential was because of its ability to withstand blade rolling and continue growing, which in turn benefited the following crop by providing weed suppression and increasing N availability (Halde and Entz, 2014). Hairy vetch's capacity to sustain productivity was also demonstrated in a 6-year organic no-till rotation where we noted an additional benefit of hairy vetch is that it is killed off during the winter and therefore did not become a weed to the following crop (Halde et al., 2015).

Despite the continuous success at the Carman Organic Crops Field Laboratory, organic farmers were, and still are, slow to adopt hairy vetch as a cover crop owing to the cost of seed, knowledge gaps in varietal choice, seeding rates, and general best management practices. To address these concerns, we investigated seeding rates for full-season hairy vetch stands and determined that hairy vetch genotypes "of northern origin" were best suited for no-till organic systems in our region (Bamford and Entz, 2017).

Green manure grazing with livestock has been documented to improve economics, mitigate greenhouse gas emissions, redistribute nutrients efficiently, reduce tillage requirements, increase microbial biomass, and the manure provided assists in soil building and fertility (Flieβbach et al., 2007; Asgedom and Kebreab, 2011; Miles and Brown, 2011; Thiessen Martens and Entz, 2011). Our research showed that in Carman, pea/oat and hairy vetch green manures were suitable forages for grazing, while soybean and lentil were not (Cicek et al., 2014). In addition, grazing the green manures provided greater benefit to the following crop versus soil incorporation (Cicek et al., 2014). When livestock grazing took place during the green manure year, net return was \$30 acre^{-1} instead of a net loss of \$168 acre^{-1}.

During the extensive investigations into green manure management and organic no-till or reduced till, in-season green manure performance was used as educational tools for farmers, researchers, and students during field days. The datasets from multiple studies were used as part of lectures in the University of Manitoba's Organic Cropping Systems course, and during lectures and seminars given at farmers' meetings and conferences in the winter months across Canada. Information from the green manure experiments at Carman were used to help develop a "Green Manure Tool" created by the POGI in 2017 (http://www.pivotandgrow.com/resources/production/green-manures/green-manure-tool-kit/). Some green manure experiments were conducted on organic farms in southern Saskatchewan and Manitoba. By partnering with organic farmers, the farmers had the ability to observe the treatment effects and learn from the research in real time. This partnership was successful as these farmers, as well as many other organic farmers we have partnered with for subsequent research, routinely engaged in conversations with graduate students running the experiments while they were taking measurements and are keen to have more research conducted on their land.

Initiating adaptive capacity

In agriculture, education to build the capacity to make informed, sound decisions in relation to static concepts is insufficient. In many ways, farmers, agronomy advisors, students, and scientists need to increase their capability to adapt to slow or sudden changes in the environment every season. The term *adaptive capacity or capabilities* is often used in a development context relating to farmers in developing countries adapting to climate change. Clay and King (2019) describe adaptive capacity as "a process that unfolds in places over time by building from recent work on adaptation pathways." The Glenlea study has served as a platform for adaptive capacity for the scientists running the experiment, the summer students working on it, and the greater community learning from it. Increasing the adaptive capabilities of organic farmers on the prairies contributes to the growth of the organic agriculture sector. Table 6.3 summarizes the main capacity building activities from the Glenlea long-term rotation study and Carman Organic Crops Field Laboratory.

The Glenlea study has hosted countless field days for farmers (2–3 per year), agronomists (local and international), and the general public and acts as a "living laboratory" each semester for a course in organic cropping systems. In addition to that, dozens of people contribute to these research projects as summer employees working at the research station.

Future work

An important aspect of long-term experiments is that they can be observed in the future for answers about problems that have yet to arise. Janzen (1995) explained it well when he wrote "Perhaps the best justification for the establishment and maintenance of long-term sites is that they provide a resource for future scientists posing questions we have not yet anticipated. There was no way of knowing in 1910, for example, that rotation ABC would one day provide information pertinent to the issue of global warming. Indeed, most of the key findings from this site could not have been envisioned at the turn of the century. Future generations of scientists, in addressing the questions that will inevitably arise regarding agricultural sustainability, will cherish the long-term ecological sites they inherit, provided they have been adequately established, documented, and maintained."

The impact of climate change on agriculture will continue to be an important area of study, and how or if organic systems are affected or could adapt differently than conventional systems could be studied at Glenlea. As an example, the first comprehensive study of nitrous oxide emissions in conventional versus organic wheat production in Canada was recently conducted at Glenlea (Westphal et al., 2018). We may see more frequent drought periods or severe weather events so being able to provide research and information to farmers to assist them in navigating a changing climate for agriculture will be important to continue producing high-quality food for the growing global population. In the Braman et al. (2016) study, the Prairie grassland soils were found to be more drought resilient than arable systems based on greater stability of microbial biomass during a summer drought. Then, when autumn rains arrived, biomass carbon recovery was greater in the organic than conventional plots. It is doubtful whether such processes would appear or be detectable in short-term studies.

Conclusion

Most of the lessons from Glenlea could only have come from a long-term experiment. Glenlea allowed problems to be uncovered and mitigation strategies to be implemented. Local and visiting researchers, together with their students, have built upon knowledge from Glenlea and created a critical knowledge bank (i.e., peer-reviewed papers and theses) and resources such as the nutrient budgeting Tool. Tools born out of the Glenlea study have informed farmers and advisors, allowing pitfalls in organic production to be avoided. Long-term studies like Glenlea provide the rare opportunity to understand the ecology as well as the agronomy of organic cropping systems, leading to a deeper understanding of sustainability. Because the Glenlea rotation study embodies a physical place, visitors can see results first hand (e.g., Figs. 7.3 and 7.6). In many ways, the Glenlea study and the Canadian organic crop production sector have grown up together, alongside each other. In this way, Glenlea is like other long-term studies where new innovations in agriculture are documented soon after their introduction. For example, in the same way that Rothamsted's Broadbalk plot continues to inform fertilizer use in wheat production, Glenlea will likely continue to inform Canada's organic crop production sector.

Acknowledgments

We first acknowledge that the Glenlea rotation study is situated on Treaty 1 territory, the ancestral lands of the Anishinabek people and the Metis Nation. We also acknowledge all who have toiled in the Glenlea plots, under windy, rainy, snowy, drought, cold, and hot conditions. We celebrate the perseverance of our technical staff, our summer students, and the farm staff at Glenlea who patiently support the research. A special thanks to the governments of Manitoba and Canada for ongoing financial support.

References

Asgedom, H., Kebreab, E., 2011. Beneficial management practices and mitigation of greenhouse gas emissions in the agriculture of the Canadian Prairie: a review. Agronomy for Sustainable Development 31 (3), 433–451.

Bachinger, J., Reckling, M., Stein-Bachinger, K., 2012. Baltic Ecological Recycling Agriculture and Society. Nitrogen Budg. Calc. Available from: http://www.beras.eu/implementation/index.php/en/software-tools. Verified 16 February 2019.

Bamford, K.C., Entz, M.H., 2017. Management of organic hairy vetch (Vicia villosa) cover crops in establishment year. Canadian Journal of Plant Science 5, 1–5. August 2016.

Blackshaw, R.E., Molnar, L.J., Larney, F.J., 2005. Fertilizer, manure and compost effects on weed growth and competition with winter wheat in western Canada. Crop Protection 24 (11), 971–980. Available from: https://doi.org/10.1016/j.cropro.2005.01.021.

Braman, S., Tenuta, M., Entz, M.H., 2016. Selected soil biological parameters measured in the 19th year of a long term organic-conventional comparison study in Canada. Agriculture, Ecosystems & Environment 233, 343–351.

Campbell, C.A., Zentner, R.P., Janzen, H.H., Bowren, K.E., 1990. Crop Rotation Studies on the Canadian Prairies, 1841/E. Agriculture and Agri-Food Canada, Ottawa, ON.

Cicek, H., Martens, J.R.T., Bamford, K.C., Entz, M.H., 2014. Forage potential of six leguminous green manures and effect of grazing on following grain crops. Renewable Agriculture and Food Systems 30 (6), 503–514.

Clay, N., King, B., 2019. Smallholders' uneven capacities to adapt to climate change amid Africa's 'green revolution': case study of Rwanda's crop intensification program. World Development 116, 1–14.

Eghball, B., Ginting, D., Gilley, J.E., 2004. Residual effects of manure and compost applications on corn production and soil properties. Agronomy Journal 96 (2), 442–447. Available from: https://doi.org/10.2134/agronj2004.4420.

Eghball, B., Wienhold, B., Gilley, J., Eigenberg, R., 2002. Mineralization of manure nutrients. Journal of Soil and Water Conservation 57 (6), 470–473. https://doi.org/10.1006/meth.2001.1262.

Entz, M.H., Guilford, R., Gulden, R., 2001. Crop yield and soil nutrient status on 14 organic farms in the eastern portion of the northern Great Plains. Canadian Journal of Plant Science 81, 351–354.

Entz, M.H., Penner, K.R., Vessey, J.K., Zelmer, C.D., Thiessen Martens, J.R., 2004. Mycorrhizal colonization of flax under long-term organic and conventional management. Canadian Journal of Plant Science 84, 1097–1099.

Entz, M.H., Welsh, C., Mellish, S., Shen, Y.Y., Braman, S., Tenuta, M., Turmel, M., Bucley, K., Bamford, K.C., Holliday, N., 2014. Glenlea organic rotation: a long-term systems analysis, pp. 215–237. In: Martin, R.C., McRae, R. (Eds.), Managing Energy, Nutrients, and Pests in Organic Field Crops. CRC Press, New York.

Evans, R., Lawley, Y., Entz, M.H., 2016. Fall-seeded cereal cover crops differ in ability to facilitate low-till organic bean (Phaseolus vulgaris) production in a short-season growing environment. Field Crops Research 191, 91–100. Available from: https://doi.org/10.1016/j.fcr.2016.02.020.

Fließbach, A., Oberholzer, H.-R., Gunst, L., Mäder, P., 2007. Soil organic matter and biological soil quality indicators after 21 years of organic and conventional farming. Agriculture, Ecosystems & Environment 118, 273–284.

Fraser, T., Lynch, D.H., Entz, M.H., Dun, K.E., 2015. Linking alkaline phosphatase activity with bacterial phoD gene abundance in soil from a long-term management trial. Geoderma 257–258, 115–122.

Grant, C.A., Monreal, M.A., Irvine, R.B., Mohr, R.M., McLaren, D.L., Khakbazan, M., 2009. Crop response to current and previous season applications of phosphorus as affected by crop sequence and tillage. Canadian Journal of Plant Science 89, 49–66. Available from: www.nrcresearchpress.com (Verified 26 February 2019).

Gruenhagen, R.D., Nalewaja, J.D., 1969. Competition between flax and wild buckwheat. Weed Science 17 (3), 380–384.

Halde, C., Bamford, K.C., Entz, M.H., 2015. Crop agronomic performance under a six-year continuous organic no-till system and other tilled and conventionally-managed systems in the northern Great Plains of Canada. Agriculture, Ecosystems & Environment 213, 121–130. Available from: https://doi.org/10.1016/j.agee.2015.07.029.

Halde, C., Entz, M.H., 2014. Flax (Linum usitatissimum L.) production system performance under organic rotational no-till and two organic tilled systems in a cool subhumid continental climate. Soil and Tillage Research 143, 145–154. Available from: https://doi.org/10.1016/j.still.2014.06.009.

Huang, X.L., Chen, Y., Shenker, M., 2005. Rapid whole-plant bioassay for phosphorus phytoavailability in soils. Plant and Soil 271 (1–2), 365–376.

Janzen, H.H., 1995. The role of long-term sites in agroecological research: a case study. Canadian Journal of Soil Science 75, 123–133.

Johnston, A.E., Pulton, P.R., 2018. The importance of long-term experiments in agriculture: their management to ensure continued crop production and soil fertility; the Rothamsted experience. European Journal of Soil Science 69, 113–125.

Jordan, N.R., Zhang, J., Huerd, S., 2000. Arbuscular-mycorrhizal fungi: potential roles in weed management. Weed Research 40, 397–410.

Knight, J.D., Buhler, R., Leeson, J.Y., Shirtliffe, S., 2010. Classification and fertility status of organically managed fields across Saskatchewan, Canada. Canadian Journal of Soil Science 90 (4), 667–678.

Lemke, R., Malhi, S., Johnson, E., Brandt, S., Zentner, R., Olfert, O., 2012. Alternative cropping systems study - Scott, Saskatchewan. Prairie Soils and Crop Journal 5, 74–84.

Li, R., Khafipour, E., Krause, D.O., Entz, M.H., de Kievit, T.R., Fernando, W.D., 2012. Pyrosequencing reveals the influence of organic and conventional farming systems on bacterial communities. PLoS One 7 (12), e51897.

Mäder, P., Fließbach, A., Dubois, D., Gunst, L., Fried, P., 2002. Soil fertility and biodiversity in organic farming. Science 296 (5573), 1694–1697.

Martin, R.C., Lynch, D.H., Frick, B., van Straaten, P., 2007. Phosphorus status on Canadian organic farms. Journal of the Science of Food and Agriculture 87, 2737–2740.

McCartney, D., Fraser, J., Minor, A., 2010. The potential role of annual forage legumes in Canada : a review. Canadian Journal of Plant Science 90, 403–420. Table 2.

Miles, R.J., Brown, J.R., 2011. The Sanborn Field Experiment: implications for long-term soil organic carbon levels. Agronomy for Sustainable Development 103, 268–278.

Nelson, A.G., Froese, J.C., Entz, M.H., 2010. Organic and conventional field crop soil and land management practices in Canada. Canadian Journal of Plant Science 90 (3), 339–343. Available from: http://pubs.aic.ca/doi/abs/10.4141/CJPS09136.

Peoples, M.B., Bowman, A.M., Gault, R.R., Herridge, D.F., McCallum, M.H., McCormick, K.M., Norton, R.M., Rochester, I.J., Scammell, G.J., Schwenke, G.D., 2001. Factors regulating the contributions of fixed nitrogen by pasture and crop legumes to different farming systems of eastern Australia. Plant and Soil 228, 29–41.

Podolsky, K., Blackshaw, R.E., Entz, M.H., 2016. A comparison of reduced tillage implements for organic wheat production in western Canada. Agronomy Journal 108 (5), 2003–2014.

Poyser, E.A., Hedlin, R.A., Ridley, A.O., 1957. The effect of farm and green manures on the fertility of blackearth-meadow clay soils. Canadian Journal of Plant Science 37, 48–57.

Thiessen Martens, J., Entz, M., 2011. Integrating green manure and grazing systems: a review. Canadian Journal of Plant Science 91 (5), 811–824. Available from: http://www.nrcresearchpress.com/doi/10.4141/cjps10177.

Thiessen Martens, J.R., Entz, M.H., Wonneck, M.D., 2015. Review: redesigning Canadian prairie cropping systems for profitability, sustainability, and resilience. Canadian Journal of Plant Science 95, 1049–1072.

Thiessen Martens, J.R., Lynch, D.H., Entz, M.H., 2019. A survey of green manure productivity on dryland organic grain farms in the eastern prairie region of Canada. Canadian Journal of Plant Science 99 (5), 772–776.

Vaisman, I., Entz, M.H., Flaten, D.N., Gulden, R.H., 2011. Blade roller-green manure interactions on nitrogen dynamics, weeds, and organic wheat. Agronomy Journal 103 (3), 879–889.

Welsh, C., Tenuta, M., Flaten, D.N., Thiessen-Martens, J.R., Entz, M.H., 2009. High yielding organic crop management decreases plant-available but not recalcitrant soil phosphorus. Agronomy Journal 101 (5), 1027–1035.

Westphal, M., Tenuta, M., Entz, M.H., 2018. Nitrous oxide emissions with organic crop production depends on fall soil moisture. Agriculture, Ecosystems & Environment 254, 41–49.

CHAPTER

Testing long-term impact of agriculture on soil and environment in Poland

7

Grzegorz Siebielec[1], Mariusz Matyka[1], Artur Łopatka[1], Radosław Kaczyński[1], Jan Kuś[1], Wiesław Oleszek[1,2]

[1]*Institute of Soil Science and Plant Cultivation, State Research Institute, Pulawy, Poland;* [2]*Biochemistry and Crop Quality, Institute of Soil Science and Plant Cultivation, State Research Institute, Pulawy, Poland*

Historical outline

Poland has a number of sites with long-term experiments located in different parts of the country and belonging to different research units (Table 7.1). The oldest long-term agriculture experiment site in Poland was established in 1921 in Skierniewice (51 degrees51′N, 20 degrees10′E) on 5 ha area with 24 fields and 588 plots. Some experiments were conducted in three and other in five repetitions. The main objective of this site was to research on the influence of organic and mineral fertilization on a number of crops, predominantly potatoes, winter wheat, and barley. The main indicator of the effectiveness was the yield, but other parameters such as physical and chemical parameters of soil, loses of nitrogen to atmosphere, and groundwater were thoroughly studied. The data obtained from this site were published in World Report No. 7 of Global Change and Terrestrial Ecosystem published in 1996 (Stepień et al., 2018).

The middle of the 20th century resulted in establishing a number of new sites with long-term experimentation. Some of them concentrated again on the combinations of manure and organic fertilization in different rotation schemes as well as the environmental consequences of such treatments. But at the same time, there was an increased interest in grain production, which resulted in an increased share of monocultures or an increased share of grain crops. A number of sites were set up where influence of monoculture versus crop rotation systems was studied. A good example of such a trend was a long-term experiment in Brody, where seven field crop rotations and monoculture of winter rye, potato, spring barley, alfalfa, oil flax, rapeseed and triticale were studied. Over a period of 60 years, plant varieties underwent changes due to the soil sickness (Blecharczyk et al., 2018).

A second major topic of this period was the new perspective to tillage systems. The studies on reduced or no tillage systems in the context of cost reduction, soil compaction problems, water retention, soil microbiology, pH, organic matter content, crop yield, and weed infestation were major research problems addressed.

Long-Term Farming Systems Research. https://doi.org/10.1016/B978-0-12-818186-7.00008-4
Copyright © 2020 Elsevier Inc. All rights reserved.

Table 7.1 Long-term experiments in Poland.

Location	Year of establishment	Subject
Skierniewice (LSU Warsaw)	1921	Interaction between mineral and organic fertilization
Mochełek (UTP Bydgoszcz)	1948	14 fertilization variants versus plant and soil characteristics
Chylice (LSU Warsaw)	1955	Production and environmental effects of long-term organic and mineral fertilization
Brody (LSU Poznań)	1957	Fertilization/crop rotation versus monoculture
Balcyny (Warmia-Mazury Univerity)	1967	Crop rotation versus monoculture
Chylice (LSU Warsaw)	1967	Zero tillage versus traditional ploughing
Grabów (IUNG)	1969	Yield in crop rotation with different cereal share
Grabów (IUNG)	1979	Organic matter balance versus properties of nonlimed light soil
Swojec (LSU Wrocław)	1990	Triticale cropping systems
Osiny (IUNG)	1994	Agricultural management in conventional, integrated, and organic farming
Brody (LSU Poznań)	1995	Seven variants of tillage systems
Malice (LSU Lublin)	1995	Potato and barley agrotechnology
Czyrna (AU Karaków)	1996	Comparison of conventional and organic farming

Zero tillage system versus traditional ploughing in relation to different crops was tested in an experiment performed in Chylice (Lenart et al., 2018).

Long-term experiments set at the end of past century focused predominantly on the comparison of different crop production systems, including conventional, integrated, and organic. High pressure on the reduction of pesticide use and consumer expectations for food free of pesticide contaminants was the driving force for an intensive research in this field. This problem was also undertaken at the Institute of Soil Science and Plant Cultivation on two experimental farms and experiments set up in 1994, which are still in progress, and some of the results will be presented in this chapter.

Long-term changes in Polish agriculture

Analyzing changes in Polish agriculture after 1960, several periods with different production dynamics and various economic and political environments can be distinguished. From 1960 to the mid-1970s, in a centrally controlled economy, rapid industrialization of the country took place. A number of large plants producing mineral fertilizers were created, and fertilization was rapidly growing, stimulating increases in cereal yields and animal production (Fig. 7.1). There was a slight decrease in the area of arable land (Fig. 7.2). Levels of wheat yields and milk production per cow were 30%–40% lower in Poland than in the neighboring Germany

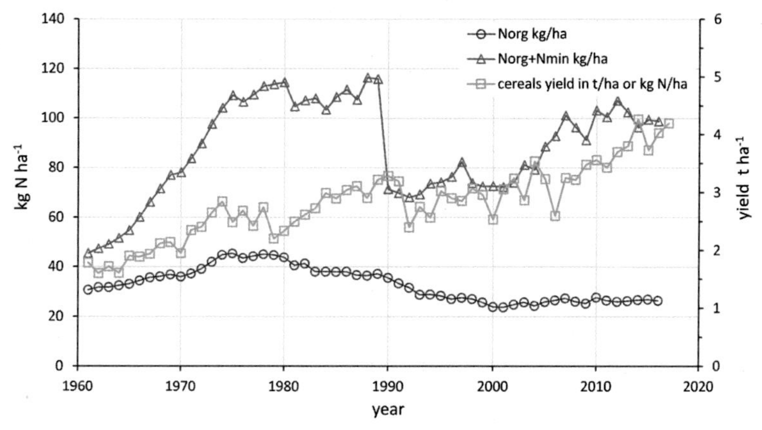

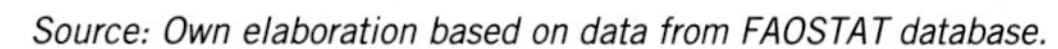

FIGURE 7.1

Mineral and organic average unit fertilizer doses and cereal yields in Poland from 1961–2016. Cereal yields (t ha^{-1}) can also be read in units of nitrogen (kg N ha^{-1}) for better visual assessment of nitrogen pollution potential (difference between triangles and squares).

Source: Own elaboration based on data from FAOSTAT database.

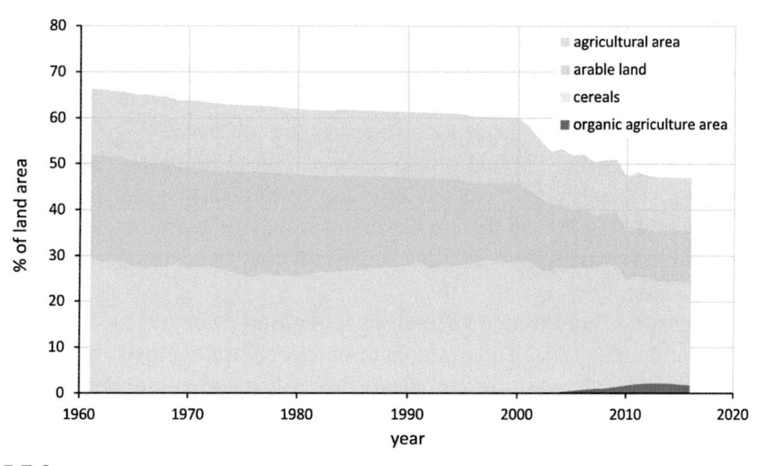

FIGURE 7.2

Land use shares by agriculture in Poland from 1961–2016.

Source: Own elaboration based on data from FAOSTAT database.

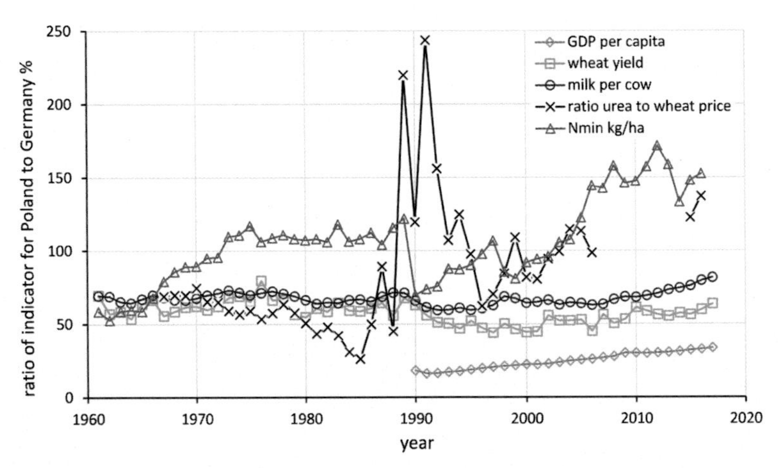

FIGURE 7.3

Polish and German agriculture 1961–2016. Curves are quotients of indicator value in Poland to the value of the same indicator in Germany.

Source: Own elaboration based on data from: FAOSTAT database (yield, milk amount, wheat and urea prices, N mineral fertilization), EUROSTAT and polish Central Statistical Office databases (urea prices), World Bank database (GDP per capita).

(Fig. 7.3). The ratio of fertilizer to grain price, which decides about crop production profitability, was more beneficial for Polish farmers. At the end of this period, nitrogen fertilization level was similar in both countries.

The next period was stagnation caused by diminishing returns from increased fertilizer application and common shortages of production factors because of wave of strikes in Polish factories. This period ended with the collapse of communism in the year 1989. The introduction of a market economy caused a short-term sharp increase in the prices of production factors, including fertilizers, resulting in a strong decline in fertilization and yield. After stagnation in the 1990s, a gentle return to previous levels of production intensity was observed (Fig. 7.1). It significantly accelerated after Poland's accession to the European Union in 2004. Unfortunately, the ratio of prices of nitrogen fertilizers to grain prices remains permanently higher in Poland than in Germany. It may be caused by better transport infrastructure in Germany, in particular, the availability of cost-effective river transport on the Rhine.

As of today, it is hard to say whether we will observe convergence of Polish and German agriculture indices. The example of nitrogen fertilizer level, which has been higher in Poland for the past years suggests that Polish agriculture has its own equilibrium level, different from the one in Germany. The reason for this may be a much higher share of sandy soils in Poland, which, among others, affects the efficiency of fertilization. It is worth to mention also that according to Eurostat, from 2005 to 2013, the average farm size in Poland increased by 69% (from 6 to 10 ha) and in Germany by 34% (from 44 to 59 ha).

In all analyzed periods, a relatively constant absolute cereal area can be observed, which, combined with the decrease in the arable land area, causes the increase in the percentage share of cereals in the structure of crops. This trend is rather negative in terms of soil quality and biodiversity aspects. Positive changes from the environmental point of view include the increase in area of organic farming observed since Poland's accession to the European Union.

Impact of crop rotation and tillage

Crop rotation and tillage are among the key elements deciding crop yields, technology intensity, and the environmental consequences of agricultural production. In recent decades, they have been subject to intense changes stimulated by biological and technical progress, availability of labor resources, and organizational or economic conditions.

Crop rotation is the central link in agricultural production, combining ecology, agrotechnology, and the tradition of cultivation of selected plant species (Niewiadomski, 1995). Continuous cropping of same crop is an extreme deviation from the rational crop rotation. However, due to economic conditions, a continuous simplification of crop rotation is observed, which can be expressed by the share of cereals in total sown area in Poland, at the level of 74%. Most of the research on succession of plants on arable fields is based on long-term observations of agrobiocenoses and edaphic factors of the habitat. The main objectives of the research on crop rotation are (Marks et al., 2018a,b)

- assessment of the importance of crop rotation in the cultivation of plants,
- assessment of the production, economic, and environmental consequences of excessive simplification of crop rotation,
- identification of edaphic factors lowering the yield of the investigated plants,
- monitoring the scale of the recruitment of cultivated species and varieties for many years of subsequent sowing,
- determination of the importance of selecting varieties, level of fertilization, and plant protection as factors compensating for continuous cropping,
- knowledge of the interaction between the occurrence of weeds, diseases and pests, and plants yielding, and
- analysis of changes in soil properties in monoculture crops against the background of crop rotation facilities.

One of the oldest experiments in Poland concerning crop rotation is the trial established in 1958 at the Brody Experimental Station belonging to the University of Life Sciences in Poznań. One of the experimental factors is the plant succession system, which compares crop rotation with a 7-year rotation of winter rye, potato, spring barley, alfalfa, oil flax, rapeseed, and triticale against the monoculture cultivation of species included in the crop rotation. In addition, the experiment included a treatment with permanent, mown fallow and black fallow as control elements to study changes in the soil treatment. In this research, it was found that in the many-year monoculture, there was a loss of organic carbon and total nitrogen in the surface layer of the soil (0–20) compared with the crop rotation with alfalfa. The greatest reduction in the content of these components was recorded on plots with continuous crops of potato and spring barley and, to a smaller extent, on winter rye. Leaving the soil without vegetation (black fallow) halved the content of organic carbon and total nitrogen in the soil, while maintaining persistent vegetation (permanent fallow) contributed to their significant increase compared with the crop rotation. Weed analysis, carried out after 50–51 years of rye cultivation in monoculture, showed a 34.5% increase in the number of weeds in comparison with the crop rotation. A definitely greater negative impact of monoculture was recorded in the assessment of weed biomass, which, in relation to the crop rotation, was almost ninefold times higher. Assessment of weed infestation of spring barley also showed a 2.5-fold increase in the number of weeds and a fourfold increase in their biomass compared with crop rotation. In addition, the cultivation of this species in the monoculture favored a much bigger infection by Gaeumannomyces graminis. On average, for the period 1958–2013, winter rye and spring barley yielded lower in the monoculture by 19.0% and 20.6%, respectively. On the other hand, potato yields were more significantly lower by 41.8% (Blecharczyk et al., 2018).

Another field experiment on crop rotation has been conducted since 1967 at the Experimental Station in Bałcyny belonging to the University of Warmia and Mazury in Olsztyn. At present, the first factor of the experiment is the succession of plants, which consists in comparing two six-field crop rotation with the monoculture cultivation of species included in crop rotation. The results of the conducted work clearly show that most species react negatively to the cultivation after each other even under conditions of intensive agricultural use such as increased fertilization, chemical protection of plants, and the use of varieties with increased resistance to diseases and pests. One of the more important results of research on monocultures is the decline effect, which consists of a significant decrease in yields in the first years of monoculture and stabilization of yields in subsequent years, but at a lower level than in the case of crops cultivated in crop rotation. It is important to note that the decreases in the yield of plants cultivated in monoculture cannot always be explained by the change in soil physical and chemical properties, the level of weed infestation, the occurrence of diseases and pests, the microbiological state of the soil, or the presence of phytotoxic substances. In this experiment, it was proved for cereals that phytocoenosis of weed plants grown in monoculture and crop rotation are characterized by a high level of floristic similarity. Multilateral crop rotation, which included species from various groups of plants (root, industrial, legumes), as well as their spring and winter forms, fulfilled a regulatory role for weeds of arable land, whereas monoculture crops promoted an increased weed density and biomass. Research on plant health has proved that cereal infection with take-all diseases (Oculimacula yallundae and G. graminis) was always higher in monoculture than in crop rotation. Deterioration of crop health under monoculture cultivation was the highest in the case of winter wheat, triticale, whereas the lowest for spring barley, and rye and oats. Taking into consideration the species of plants in the initial period of research, flax and potato reacted the most to continuous cultivation. In the last series of experiments, pea, horse bean, and sugar beet were the most susceptible for sowing after each other. Generally, cereals were less sensitive to cultivation in monoculture, but they were not a homogeneous group. Definitely the smallest differences in productivity were obtained in the continuous cultivation of winter oilseed rape and maize. In addition, it was shown that monoculture cultivation supported by mineral and organic fertilization does not cause the loss of organic matter or reduction of the content of main macronutrients. However, it has a negative effect on the pH of soil (Marks et al., 2018a,b).

Another experiment initiated in 1969 at the Experimental Station Grabów, belonging to the Institute of Soil Science and Plant Cultivation—State Research Institute in Puławy, evaluates the impact of the long term use of 4-year rotations of cereals on the yield and selected chemical properties of the soil. The crop rotation differed in a varied proportion of cereals, which was A-50%, B-75%, C-75%, and D-100%. Also, the possibility of compensating an unfavorable position for cereals through fertilization, chemical protection of plants, and the selection of different varieties was also examined. The obtained results allow concluding that in the case of high cereal share in crop rotation (75% and 100%), relatively large and stable crops can be obtained when appropriate technology solutions are implemented. On average, for the years of 2001–16, they amounted to 5.9 tonne ha^{-1} in crop rotation composed of only cereals, 6.3–6.6 tonne ha^{-1} in a crop rotation containing 75% of cereals, being smaller by 11%–20% compared with the crop rotation with a 50% share of cereals. Reductions in cereal yields due to the increase in their share in crop rotation were at a certain level in the first years and did not worsen during the experiment. Modification of the production practices (fertilization, chemical protection of plants, selection of varieties) resulted in a significant increase in yields for winter cereals, while stagnating them for spring cereals. Weeds were the decisive factor lowering yields only in the first years of conducting the experiment, when winter cereals dominated and selective herbicides were not used. In the following years, the use of correct agrotechnology and appropriately selected herbicides enabled an effective reduction of weeds, regardless of the type of cereals species, forecrop, and crop rotation. It was also found that the long-term use of crop rotation did not cause a decline in soil fertility, whereas the cereal monoculture has worsened some soil parameters (Kuś and Smagacz, 2018).

Summing up the long-term experiments on crop rotation in Poland, it can be concluded that with the simplification of crop rotation, the yields of most crops are reduced, but to a different extent, depending on the species. The highest decreases in yields were observed in the initial period of simplification of crop rotation, and in subsequent years, their stabilization was noted. The decrease in yields was generally associated with changes in weed infestation, infection of plants by diseases and pests, and deterioration of selected soil parameters.

The rational construction of crop rotation is also a part of the idea of sustainable development of agriculture, which should enable to simultaneously achieve economic and environmental goals. An important element in this aspect is also the need to limit interference in the soil system, as exemplified by simplifications in tillage, strip till, or in extreme cases, direct sowing. Plow cultivation is the oldest and most developed soil tillage system that will enable good soil preparation for sowing and plant growth, but it has many disadvantages. A change in soil cultivation technology involves many changes in the soil environment, which can be seen immediately after the modification is introduced or only become apparent over a longer period of time. The soil tillage system also affects directly and indirectly the yield level of cultivated plants (Małecka-Jankowiak et al., 2018). Therefore, conducting long-term field experiments in this area is fully justified, and their main goals are

- assessment of production, economic energy, and environmental effects of changes in the soil cultivation technique,
- analysis of changes in soil properties,
- assessment of the effect of soil cultivation techniques on weed infestation and infection by pathogens of cultivated plants,
- assessment of changes in greenhouse gas emissions as a result of modification of soil cultivation techniques.

Research on direct sowing has been carried out since 1975 at Experimental Station Chylice belonging to the Warsaw University of Life Science. The purpose of the experiment is to determine the long-term abandonment of soil cultivation on crop yields, weed infestation, as well as physical and chemical properties of the soil. For this purpose, the traditional plow cultivation with direct sowing was compared as the first research factor, wherea sthe second factor was the use of herbicides or their absence (control). In the first years of the experiment (1976–80), yields of crops were not significantly diversified, but since 1981, on the plots without herbicides, when using direct sowing, significant yield decreases and an important increase in weeds was observed. Therefore, in the following years, the scheme was changed, and herbicides were applied to all the treatments. It was found that discontinuation of traditional cultivation and introduction of direct sowing resulted in an increase in the number and mass of weeds in total, initially in relation to annual, and later multiannual species. Both in the case of winter oilseed rape and spring wheat, the abandonment of soil cultivation resulted in an increase in weed infestation and a significant decrease in yields compared with the treatments with traditional cultivation. However, differences in yields were highly dependent on the course of weather, and therefore, there were also years when these differences were not recorded. The studies have also shown that the long-term direct sowing, compared with plowing, does not significantly reduce soil productivity but protects the soil against the loss of organic matter. As a consequence, it promotes the development of soil fauna and soil aggregation. Moreover, it increases soil resistance to blurring, the field water capacity, and soil moisture (Lenart et al., 2018).

Experiments on simplification in tillage have also been conducted at the Experimental Farm Brody belonging to the University of Life Sciences in Poznań, since 1995. The aims of the experiment are both the comparison of traditional plow cultivation with direct sowing and the evaluation of a four-field crop rotation against wheat cultivation in monoculture. In addition, in 1999, a field experiment, which determines the impact of periodic interruption of direct sowing by plowing or shallow cultivation tillage, was established. In the first experiment, it was found that plants responded to direct sowing in different ways. Wheat, independently of the fore crop, yielded lower in direct sowing compared with the traditional tillage. It should be noted that the highest yield decrease was recorded in the treatment with direct sowing and cultivation of wheat in monoculture. On the other hand, the cultivation system had a small influence on yields of spring barley and pea. In the studies on the frequency of direct sowing, in the initial period, smaller yields of winter wheat were observed in simplified tillage, and especially in direct sowing, than in the traditional tillage. Different frequency of interrupting direct sowing by another tillage technique did not affect the level of yields obtained. In the following years, a differentiated reaction of winter wheat to the tillage technique was noticed, as the yields varied depending on the course of atmospheric conditions. On average for the years 2000–07, the yield of winter wheat was only slightly higher in plow cultivation against the background of simplified tillage, whereas this species yielded the lowest in direct sowing used permanently and periodically interrupted with 1-year tillage. With regard to soil properties, the accumulation of organic carbon in the 0–10 cm and decrease in the 10–20 cm layer was observed. Similar relationships were noted for total nitrogen and C:N ratio. The influence of simplified tillage, especially direct sowing, on the increase of dehydrogenase activity in 0–10 and 10–20 cm layers, and acid phosphatase only in the 0–10 cm layer as compared with traditional cultivation, was also demonstrated. After 11–15 years of using different tillage systems, the lowest cereal weed infestation was noted, expressed by the number and fresh weight of weeds in direct sowing, while the largest in the simplified tillage (Małecka-Jankowiak et al., 2018).

Summing up, the long-term experiments on tillage in Poland showed that along with the simplification of soil tillage, a decrease in yields of most crops is observed, but the species reaction is diversified. The highest yield decreases were observed in the initial period of tillage simplification, while in the subsequent years, stabilization was observed. It should be added that in the subsequent years of conducting experiments, a significant influence of weather conditions on the differences in yields between different tillage systems is observed. Simplifications in tillage generally increase the number and weight of weeds, but the selected soil parameters improve, for example, organic matter content and microbiological activity.

Impact of crop production systems

Since 1994, in the Experimental Station in Osiny (51 degrees28′ N, 22 degrees03′ E, 155 MAMSL) belonging to IUNG-PIB Puławy (Institute of Soil Science and Plant Cultivation—State Research Institute), there has been an ongoing experiment in which different crop production systems are compared. The experiment was established on the podzolic soil with granulometric composition of loamy sand, passing at a depth of 60–70 cm into sandy loam. On a small part of the field, there is also black soil of similar texture. The soil was slightly acidic (pH in KCl 4.4–5.4), the average abundance in phosphorus and potassium and the organic carbon content in the range of 0.78%–0.84%. The experimental field with a total area of about 16 ha has been divided into parts representing the compared crop production systems (Fig. 7.4). The area of each field is about 1 ha, which allows testing effects of a given practice close to the production conditions. In each field, several varieties are sown, which allows to additionally determine their suitability for various production systems (Kuś et al., 2010).

In this experiment, the following production systems are tested:

I. Ecological system represents a five-field crop rotation: potato-spring barley (from 2005 spring wheat) + undersown (a mixture of red clover with white clover and grasses)—clover grass used 2 years—winter wheat + intercrop. Organic fertilization includes the use of 25–30 tonne ha^{-1} compost for potatoes and plowed catch crop (mixture with the participation of Fabaceae). Since 2002, potassium fertilization has been at a dose 40 kg ha^{-1} K_2O and, since 2007, phosphorus has been at a dose 15 kg ha^{-1} P_2O_5 in fertilizers allowed for organic farming. Regulation of weed infestation consists of intensive mechanical procedures, and additionally, hand-weeding of the potato before the last hilling. In potato, NOVODOR, based on *Bacillus thuringiensis* subsp. *tenebrionis*, is used for controlling Colorado beetle and copper preparations for limiting the development of potato blight.

II. Integrated system includes the following crop rotation: potato-spring barley (from 2005 spring wheat) + intercrop—seed bean—winter wheat + intercrop with cruciferous plants.

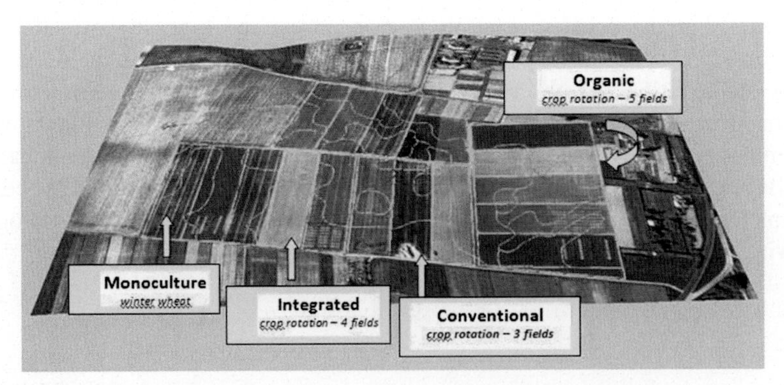

FIGURE 7.4

Experimental field at Osiny experimental station IUNG-PIB near Puławy.

Organic fertilization includes 25–30 tonne ha^{-1} of compost for potatoes, plowed horse bean straw, and cruciferous plants intercrops. Doses of phosphate and potassium fertilizers balance their take-off with yields of plants, and nitrogen fertilizer doses are adjusted based on the determination of the amount of mineral nitrogen forms in the soil layer 0–90 cm before spring vegetation, additionally taking into account the state of wheat nutrition with nitrogen (SPAD). In total, fertilizer doses are 30%–40% lower than in the conventional system. The use of chemical plant protection treatments depends on the intensity of individual pests.

III. Conventional high-volume system occurring in two variants A and B.

Variant A includes a rotation of three-field crops: rapeseed, winter wheat, and spring barley (from 2005, spring wheat)

Management in this system is based on intensive cultivation technologies recommended by the IUNG, which are characterized by high consumption of industrial means of production, and organic fertilization is limited to plowing straw rape and winter wheat.

Variant B includes monoculture of winter wheat

In this variant, intensive production technology is used, aimed at the effective reduction of pests and good supply of nutrients to the plants. Organic fertilization was in the form of straw plowed every other year.

Plant yields

Winter wheat yielded the highest in the integrated system, 6.7 tonne ha^{-1} on average for 20 years (1996–2015), where it was sown after the bean plants under conditions of limited consumption of industrial means of production. A similar yield of winter wheat was obtained after rapeseed in the conventional system; however, a higher nitrogen dose and more chemical plant protection treatments were used here. In monoculture (continuous cultivation of winter wheat since 1992), despite the use of intensive production technology, grain yield on average for 20 years was by 1.6 tonne ha^{-1} (24%) smaller than in the integrated system. In the ecological system, the yield of winter wheat sown after clover with grass used for 2 years, on average for 20 years, was lower by 2.3 tonne ha^{-1} (34%) than in the integrated system.

In the integrated system, there was a clear trend of grain yield growth in the analyzed 20 years. A certain tendency of its growth was also noted in monoculture, whereas in the ecological system, there was no clear trend of productivity growth, whereas large fluctuations in yields occurred in years.

The lower yield of wheat in the monoculture and ecological system was a consequence of by about 20% smaller number of spikes and a lower weight of 1000 grains (Kuś et al., 2010; Sadowski et al., 2010).

The spring barley, on average for 9 years (1996–2004), yielded on a similar level in the system integrated after potatoes fertilized with compost, under conditions of using medium-intensive production technology and in a conventional system in winter wheat stand, where intensive production technology was applied. In the ecological system, also after potatoes, its yield was by 15% lower. Yield of spring wheat, on average for 11 years (2005–15), was the largest (5.4 tonne ha^{-1}) after potato in an integrated system. In the conventional system, where spring wheat was sown after winter wheat, despite the use of intensive production technology, its yield was lower by an average of 10%. In organic farming, its average yield was by 40% lower than in the integrated system. It should be emphasized that in the organic cultivation, a mixture of clovers and grasses was undersown into the spring cereals. A successful undersown crop of grass and clover could also compete with cereals for water and nutrients, which further reduced their yield.

The average potato yield for 20 years in organic farming was 32% lower than in the integrated system, and in the years the reduction, it ranged from 13% to 66%. The extent of yield loss in organic system depended mainly on the date of occurrence of potato blight (*Phytophthora infestans*) and the rate of its development conditioned by the course of the weather, as the effectiveness of copper preparations under conditions of high disease pressure was limited. In the ecological system, in addition to the lower yield, the share of large bulbs constituting marketable yield was significantly smaller.

Among the compared plant species, the highest productivity expressed in units of cereal grain equivalent was recorded for clover grass—108 grain units (59 in both the first and in the second year of use) in the organic system. As a result, the total efficiency of the ecological and integrated system was similar, whereas the total efficiency of the conventional system was lowered by 10%, and of winter wheat monocultures by 24%.

Weeds

A detailed analysis of weed infestation included the number and species composition and weed biomass in all cultivated plant species in at least two development phases. The dry mass of weeds in the field of winter wheat before its harvest in the ecological system was on average around 80 g m^{-2}. In most years, it did not exceed 50 g m^{-2}, and only in 3 years, it reached 200 g m^{-2}, which was caused by the lower density of wheat due to worse emergence or weaker wintering.

In spring cereals in organic farming, the dry mass of weeds was on average around 20 g m^{-2}. In this case, a successful undersown crop of clovers with grasses constituted the competition for weeds and limited their growth.

At all other facilities, the used herbicides effectively limited weeds.

Soil properties

The organic carbon content in the soil in the ecological system in the years 1996–2015 increased only by 0.03%–0.04%. This increase was smaller than one calculated according to the VDLUFA methodology (VDLUFA, 2014), which could have been caused by a large number of mechanical cultivation and care treatments used to reduce weed infestation. On other treatments, the organic carbon content in the soil was slightly lower than in the initial state.

The potassium content of the soil after 5–6 years of ecological management, decreased to a very low level, and symptoms of deficiency of this component appeared on the potato. Detailed studies also showed insufficient nutrition of winter wheat with potassium. The negative balance of potassium was about 130 kg ha $year^{-1}$ in this system, and the large yields of clovers with grasses were of decisive importance, as more than 500 kg ha^{-1} of this nutrient was utilized by plants for the period of 2-year use. After the introduction of mineral potassium fertilization since 2002, in an annual average dose of 33 kg K ha^{-1}, the soil potassium has stabilized at a low level.

In the case of phosphorus, the soil in ecological system also showed a tendency of depletion after 8–9 years. This trend was corrected with the introduction of fertilization with ground phosphate rock (7 kg P ha^{-1}) since 2007. There was no deficiency in the nutritional status of plants with phosphorus, and the factor improving the circulation of this component may also be the high biological activity of the soil, and particularly, the high activity of phosphatases, such as acid and alkaline, which was found in this treatment (Martyniuk et al., 2007; Stalenga, 2007).

Indicators characterizing biological soil activity (biomass of microorganisms, number of bacteria, soil respiration, dehydrogenase activity, acid and alkaline phosphatase activity) in the ecological system were much higher than in other treatments. Ecological management also increased the population of soil microorganisms that are involved in nitrogen transformation: N_2 assimilators of the genus *Azotobacter* and symbiotic bacteria, ammonifiers, and nitrifiers (Martyniuk et al., 2007).

In addition, the ecological management system increased the share of large soil aggregates and significantly improved their durability and, as a consequence, the share of large pores in the structure of soil pores. Soil infiltration capacity was also significantly better under ecological management (Król et al., 2012).

Long-term impact of agriculture at regional level on soil organic carbon accumulation

There is an increasing demand for evaluating the impact of organizational and economic factors in agriculture on soil carbon balance. This problem was also undertaken at the Institute of Soil Science and Plant Cultivation by evaluation of the effects of organizational and economic factors in agriculture on organic carbon content changes in soils of arable land at regional scale, in a few selected Polish regions. The specific objective of this study is to determine the magnitude and direction of changes in organic matter content in the soils in relation to the dynamics of long-term changes taking place in agriculture.

The starting point for the analysis was a database of reference soil profiles described and analyzed during the compilation of the national soil agronomic map 1: 25,000 in the years 1960–84. All profiles have been digitalized and contain results for soil horizons up to a depth of 1.5 m for each profile. A set of information describing profiles contains land use, location of the profile within the landscape, soil complex, soil type, texture, organic matter content, pH, and available nutrients. This database was developed at the IUNG, Puławy. Many properties of soils recorded in this database are now out-of-date, but the database itself is a reference point for investigations concerning changes in soil properties (for example, SOM changes) and relationships between various soil properties.

In the case study of Dolnośląskie region (Fig. 7.5), the impact of agricultural production systems on SOM has been estimated by linking changes in the content of C-org with organizational changes in agriculture over a long time period 1960–2010. To evaluate soil organic carbon (SOC) content changes at regional scale, we use measurements from a soil inventory dating from 1960 to 80 and resampled at 2003–13. Resampling of soil profiles over a long time period in

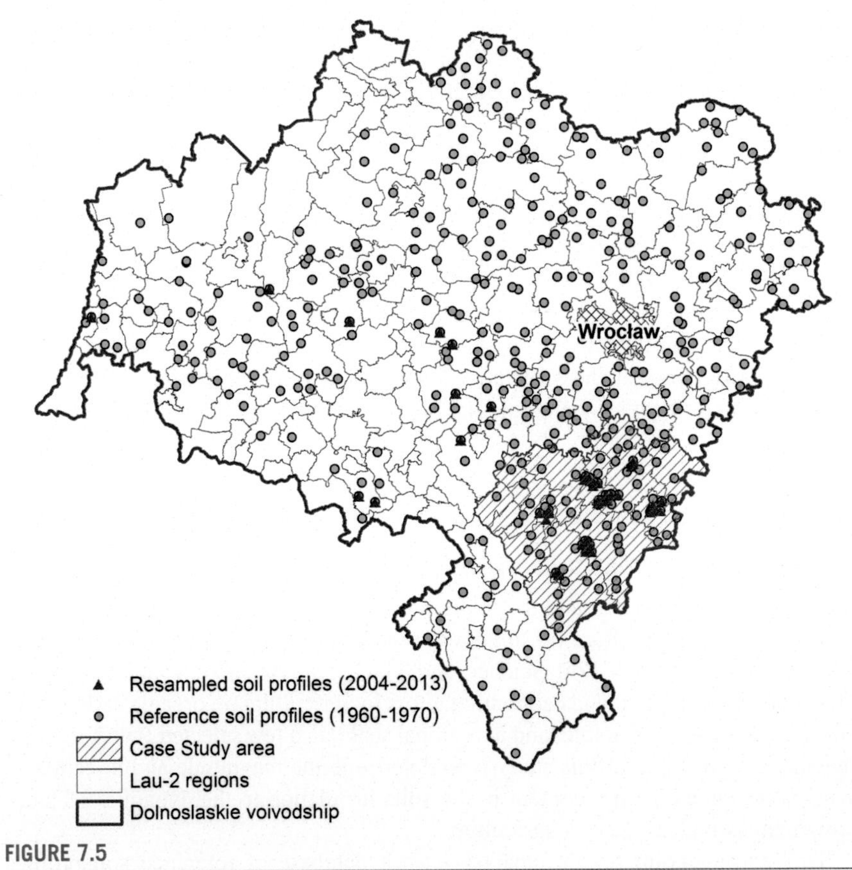

FIGURE 7.5

Location of reference soil profiles in the case study area.

combination with relatively detailed information about agronomy from 1960 up to now probably allows detecting significant changes in SOC content in response to agricultural management.

The data about agricultural management were derived from Central Statistical Office, extracted from National Agricultural Censuses. The data include farms structure, livestock density, crop rotation, yields, and amount of natural and mineral fertilization from 1960 to 2010 period in the 8–10 years series. The study was performed based on a reexamination of soil samples from the same places while maintaining long-term time interval and combining georeferenced soil data with agricultural survey data in the sense of organizational and economic factors and their changes over decades.

Production direction in that period has changed considerably from the mixed cropping–animal farming to highly specialized crop production without livestock, the so-called “cash cropping.” The agriculture in the CS Dolnośląskie (the southeast part of that voivodship) is very intensive, characterized by a large use of agricultural inputs and by a very low livestock density. The farms in Dolnośląskie are relatively large, compared with the national average, with the mean area of farm about 19 ha (taking into account only farms > 1 ha UAA) and 32 ha (only farms > 5ha UAA).

The RothC was applied to model SOC changes in the study area, as resulting from agronomy and pedoclimatic conditions. It is one of the best known models used to study the dynamics of carbon in soils, also as a response to changing climatic and agronomic/habitat conditions. The model was run for 94 locations with the known initial (in 1960s) SOC content across the study area. The step approach was applied for modeling soil C using 10-year steps, corresponding to the collection of official agricultural statistics. The soil C result of each modeling step constituted an input data to the subsequent step. Such approach allows a better illustrating of the trend of changes in agronomic factors and their impact on SOC stock changes.

Both the modeling results and comparative soil analysis in the same locations revealed similar observations. Current intensity of crop production and the related level of plant residues left in the soil causes SOC accumulation process in many soil locations. In the 1960–2014 period, SOC accumulation was observed in almost all locations of the area characterized by low initial SOC content, but this trend was not visible in soils with high initial SOC content. An analysis of modeling results for individual locations revealed that the result was largely dependent on the initial SOC content. Soils, initially very low in C, exhibited a constant accumulation of C within the whole 1971–2014 period, with a higher intensity of the accumulation since the early 1990s (Kaczyński et al., 2016).

Impact of exogenous organic matter application on soil and water quality

Availability of exogenous organic matter

Due to the permanent deficit of manure in many regions of Poland, considered as the main source of exogenous organic matter (EOM) in the past, commonly available alternative sources of organic matter such as sewage sludge, municipal biodegradable waste converted to compost or digestate, are taken into consideration. Besides sewage sludge, the current national legislation allows using other EOM sources in agriculture: e.g., substrate from mushroom production, fish pond sediment, crop pomace, and pulp. The amount of waste from the production of mushrooms is at the level of 1500 1000 tonnes, constituting a significant potential in the field of soil fertilization. These wastes can contain about 30% organic carbon, 2% nitrogen, and 0.5% phosphorus. The current amount of biodegradable waste from municipal green areas is around 340,000 tonnes per year, constituting a substantial resource of raw material for composting, and according to forecasts, it will not undergo any drastic changes in the coming years.

One of major sources of EOM is municipal sewage sludge. The intensive development of the economy, including the modernization and construction of new sewage treatment plants, as well as the extension of the sewerage network, resulted in increased production of sewage sludge in Poland. Statistical data have indicated municipal sludge production in Poland at the level of 500–600,000 tonnes of dry matter per year during the past decade with a slight increase observed (Fig. 7.6). The forecasts assumed that in 2018, this number could have reached as much as 706,000 tonnes; however, the data is not available (Bień, 2012; GUS, 2018). The amount of sludge used in agriculture has remained at the level of 100,000 tonnes since 2007. This means that share of sludge utilized as a source of organic matter in agriculture is relatively low.

The quality of municipal sewage sludge used in agriculture is specified in the Ordinance of the Minister of the Environment of February 6, 2015. The criteria

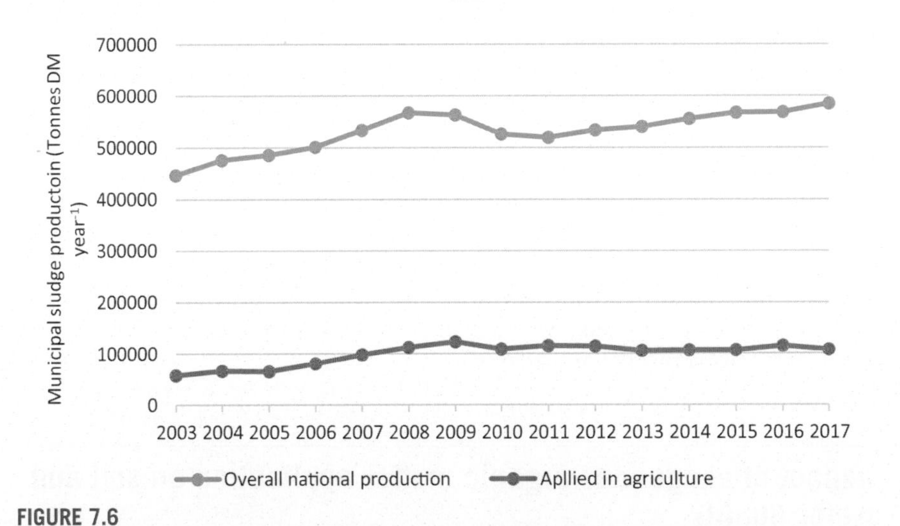

FIGURE 7.6

Annual production of municipal sewage sludge in Poland and its application in agriculture.

include quality indicators, such as the content of trace elements (cadmium, lead, mercury, nickel, zinc, copper, chromium), the presence of Salmonella, and the number of live eggs of intestinal parasites (Ascaris spp., Trichuris spp., Toxocara spp.). The regulation also specifies permissible annual or cumulative 2- to 3-year doses of sludge used in agriculture and reclamation for nonagricultural purposes. Despite the fact that the sludge criteria and application conditions are regulated in detail, there are still many controversies around the use of sludge as soil amendment. They result mainly from the perceived risks related to presence of biological and chemical contaminants in the sludge and the complicated permission procedures or procedures converting sludge into a certified soil improver.

Sludge characteristics

A nationwide survey of municipal sewage sludge quality has been done by Siebielec and Stuczynski (2008), who aimed to recognize a level of metals and nutrients in the sludge presently generated in Poland. Metal contents were assessed in relation to current thresholds and potential risk related to biosolids introduction to soil environment. The study covered a group of 60 biosolids representing different technologies of water cleaning, municipality size, and geographic location. For this survey, 31 biosolids were collected in industrial Upper Silesia region, whereas 29 samples representing other areas of Poland served as a reference group. Total contents of trace metals and other macroelements were determined by atomic absorption spectrometry or atomic emission spectrometry after hot aqua regia digestion procedure. Among trace metals, cadmium (Cd), chromium (Cr), and nickel (Ni) contents were the most diversified. However, 68% of all sludges met all trace metal content criteria for the use in agriculture. Most frequently, threshold values were exceeded by zinc, the element that in fact is generally not toxic to humans, through any of the soil-containing exposure pathways. Thresholds for most toxic metals, such as Cd and Pb, were met by 93% and 97% of all tested sludges. Interestingly, there were no differences between sludges from the postindustrial region and the reference group in terms of meeting the criterion of metal content.

The study revealed that, on average, municipal sludges produced in Poland contain 2.6% of nitrogen (N) and 1.83% of phosphorus (P) in dry matter (Siebielec and Stuczynski, 2008).

Assuming the forecast sludge production at the level of 706.6 thousand tonnes, at national level, it contains almost 18.5 thousand tonnes of N and 13 thousand tonnes of P. For example, this amount of P would be sufficient to replace mineral P fertilizers at the area of 618 thousand hectares of arable land (6.2% of total arable land) annually, since the average application rate of P in Poland is 20.9 kg ha^{-1} (Siebielec et al., 2018).

Impact of sewage sludge on crop yield, soil chemical, and biological processes

The influence of high rates of sewage sludge application to the soil on crops, biological properties of soil, and groundwater quality was tested in a long-term experiment using concrete lysimeter plots, filled with loamy sand and sandy loam in the topsoil and subsoil, respectively. The experiment was carried out using 24 lysimeters of 1 m^2 area, located in Pulawy, Poland, described in detail in Siebielec et al. (2018). The soil amendment variants are provided in Table 7.2.

In 2006, in four variants, sludge (SL) was used at rate equivalent to 100 tonne sludge dry weight per hectare. After 6 years, on selected plots, sewage sludge was applied again to obtain various variants, corresponding to a single dose of 100 tonne of sludge with varying time from application or a cumulative dose of 200 tonne ha^{-1}.

To compare sludge effects with other EOM, solid digestate (DG) was applied after prior mineral nitrogen (MN) fertilization or cumulated with prior application of sewage sludge. The effects of these EOM amendments were confronted with results collected for the unamended soil (CTRL) and with constant fertilization with mineral N. In addition, since 2012, one plot fertilized with sludge in 2006 has been left uncultivated as a fallow (FA) and not fertilized to record a barren soil effect after sludge application.

In the years 2012–15, the crop rotation was as the following: maize, winter wheat, spring barley, winter rape, and local crop cultivars. Mineral nitrogen was applied as ammonia at rates equivalent to 200, 150, 150, and 150 kg N per hectare for maize, wheat, barley, and rape, respectively. Phosphorus was applied to all plots as superphosphate 46% at 35 kg P_2O_5 ha^{-1} rate, for each crop, while potassium also was applied to all plots as potassium chloride fertilizer (90 kg K_2O ha^{-1} for each crop). Both in 2006 and in 2012, sewage sludge of earthy consistency, produced in the nearby Nałeczów wastewater treatment plant, was applied and mixed thoroughly with 0–15 cm soil layer. The trace element content did not exceed the permissible content for the use of municipal sewage sludge in agriculture, in accordance with the Ordinance of the Minister of the Environment of February 6, 2015, on municipal sewage sludge. The solid digestate was produced in an agricultural biogas plant using corn silage, grass silage, and stillage as feedstock.

Table 7.2 Fertilization regime in the lysimeter plot experiment (Siebielec et al., 2018).

Soil treatment	Years 2006–11	Years 2012–15
CTRL	No fertilization	No fertilization
SL–MN	Sludge of 100 tonne ha^{-1}	Mineral N
SL–DG	Sludge of 100 tonne ha^{-1}	Digestate 50 tonne ha^{-1}
SL–SL	Sludge of 100 tonne ha^{-1}	Sludge of 100 tonne ha^{-1}
SL–FA	Sludge of 100 tonne ha^{-1}	No fertilization, fallow
MN–MN	Mineral N	Mineral N
MN–DG	Mineral N	Digestate 50 tonne ha^{-1}
MN–SL	Mineral N	Sludge 100 tonne ha^{-1}

Organic fertilization applied in 2006 or 2012, mineral N applied every year.

The fertilization scheme had a statistically significant effect on the SOC content. The highest SOC was found on the plots where the organic material (SL-SL, DG-SL) was applied twice, where SOC increased from 10.2 g kg^{-1} in CTRL to 13.5 and 14.3 g kg^{-1}, respectively, and on the fallow plot where sludge had been applied in the first period (13.9 g kg^{-1}). Higher SOC in soil fertilized with MN than in the CTRL soil results from higher crop residue inputs, from both shoots and roots, in the ammonium nitrate variant. The fertilization variant significantly affected P availability—all sludge treatments resulted in four to six times higher content of available P as compared with permanent mineral fertilization. As mentioned before, sludges are substantial source of P; therefore, the availability of soil P originating from sludge amendment increases over time due as a result of organic P mineralization (Mackay et al., 2017).

A considerable increase in the shoot and grain yields of corn, wheat, barley, and rape was observed in the sludge-treated soils as compared with CTRL soil. For corn, grown in the first year after sludge application in 2012, the increase was the most pronounced among all crops: 1.8–2.4 times higher straw and 4.2–5.2 higher grain yield in various sludge treatments compared with CTRL soil. The highest corn yields were driven by the following sludge treatments: SL-DG treatment for shoot biomass (22,400 kg ha^{-1}), whereas SL-MN, SL-SL, and MN-SL treatments for grains (13,900–14,000 kg ha^{-1}). These sludge effects were comparable with mineral N fertilization (MN–MN) that also resulted in high corn yields (Table 7.3).

Across all crops, all the sludge variants enabled high shoot and grain yields at similar level as MN–MN fertilization, regardless of total sludge rate and time of application.

In our study, the digestate applied after prior mineral fertilization promoted high corn and wheat yields but did not increase the yields of barley and even diminished the yield of rape in 2015 in comparison with CTRL. The weaker effect of DG in every subsequent year of the experiment likely resulted from depletion of nitrogen.

The study revealed that the sludge application also promoted the abundance of culturable soil bacteria, even when applied at the reclamation rates. Bacteria, both total and those driving ammonification processes, were the most abundant in the soils fertilized with sewage sludge. In the case of all measured soil enzymes (alkaline and acidic phosphatases and dehydrogenases), the highest activity was found for the cumulative dose of sewage sludge—200 tonne ha^{-1} within 6 years. For example, dehydrogenase activity was the highest in SL-SL treatment, followed by SL-DG and MN-DG. The lowest activity was observed for MN–MN and the unamended control soil (Fig. 7.7). The study revealed that the sludges that meet the quality criteria can be applied in agriculture without negative effects for soil microorganisms or the microbially mediated soil processes.

However, the studies on the long-term sludge impact on the changes in structure of microbial communities are also needed to explain its effects on microbial diversity.

Table 7.3 Straw and grain yields (% of CTRL) in years 2012–15 (Siebielec et al., 2018).

	Corn		Wheat		Barley		Rape	
Soil treatment	**Straw**	**Grain**	**Straw**	**Grain**	**Straw**	**Grain**	**Straw**	**Grain**
CTRL	100 d[1]	100 c	100 b	100 b	100 b	100 b	100 c	100 ab
SL–MN	220 ab	523 a	113 ab	125 a	127 a	119 ab	138 a	145 a
SL–DG	238 a	415 b	107 b	105 b	114 ab	112 b	116 bc	112 ab
SL–SL	189 bc	521 a	127 ab	129 a	126 a	124 a	145 a	150 a
SL–FA	–	–	–	–	–	–	–	–
MN–MN	226 ab	523 a	123 ab	128 a	125 a	123 a	117 bc	153 a
MN–DG	200 abc	415 b	124 ab	119 ab	110 b	106 b	88 c	79 b
MN–SL	176 c	521 a	134 a	138 a	129 a	130 a	123 ab	126 ab

Means followed by the same letters are not significantly different between the inoculation types according to the Tukey's test with P < .05.

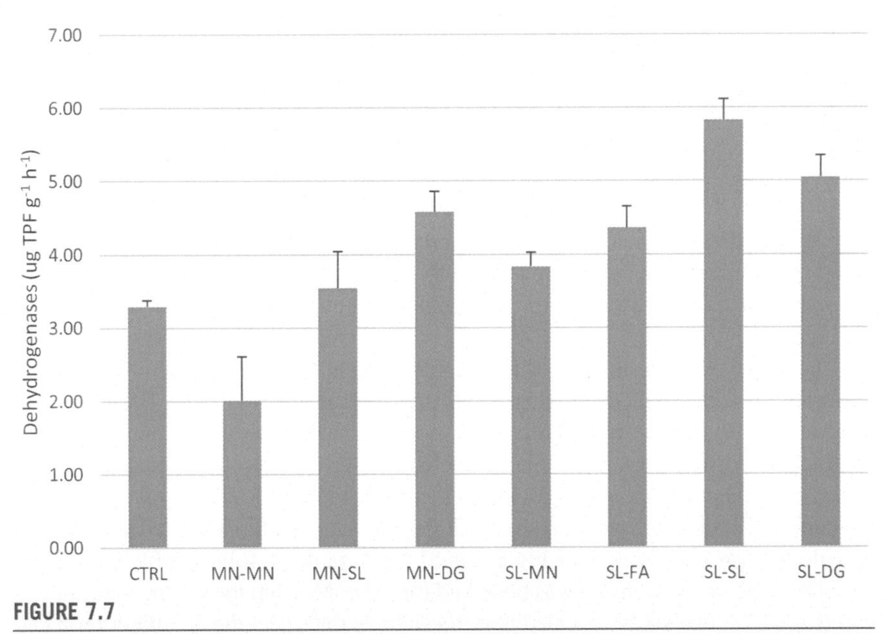

FIGURE 7.7

Dehydrogenases activity resulting from soil amendment with various sludge application variants (standard deviation presented with whiskers).

Impact of sewage sludge on groundwater quality

Lysimeter leachates were collected for three seasons: 2011/2012, 2012/2013, and 2013/2014, and leachate samples were averaged to express annual leaching of metals and nutrients over 12 months from November to October of the following year. The total amount of leachate collected from lysimeters differed considerably in individual years, while there were no substantial differences in leachate volume between soil amendment variants.

The average concentration of nitrates in leachates varied throughout the years. The lowest concentrations were recorded in the 2012/2013 season, whereas higher concentrations recorded in 2011/2012 and 2013/2014. Irrespective of the season, the highest concentrations of nitrates were recorded for the barren soil applied after prior soil treatment with sludge (SL-FA). It resulted from the lack of crops, which made the nitrogen released through slow mineralization of sludge to be leached through the soil profile. The lowest concentrations of nitrates were found in leachates from the control plots and those fertilized with digestate. In the 2011/2012 season, the nitrate concentration, in general, was at the level characteristic for class II of groundwater purity (10–25 mg $N-NO_3$ L^{-1}) according to the Regulation of Ministry of Environment from 2004. In the 2012/2013 period, the level of 10 mg $N-NO_3$ L^{-1} was exceeded only for SL-FA, whereas in the 2013/2014 season, some leachates reached class I of groundwater quality, except for SL-FA, SL-SL, SL-MN, and MN-SL. Class I is described as most clean water, without anthropogenic impacts. In the case of phosphates, the concentrations were low, at or below the detection limit for all plots.

The lowest total nitrogen leaching from the soil profile in 2011/2012 season resulted mainly from the small amount of leachate collected. The N leaching increased in subsequent periods (Fig. 7.8). In the season 2013/2014, N leached from SL-FA reached the amount corresponding to almost 60 kg ha^{-1}. Such fertilization variants as SL-MN, SL-SL, and MN-SL were in the range of 30–35 kg N ha^{-1}. In other combinations, the nitrogen leaching did not exceed 20 kg ha^{-1}.

The concentrations of potentially toxic trace elements: arsenic (As), Cd, and Pb in the leachates underwent minor changes in the particular seasons and, interestingly, were at a level many times lower than the permissible limits for drinking water, specified in the Ordinance of the Minister of Health from March 29, 2007 on the quality of water for human consumption. The threshold contents for As, Cd, and Pb are 10, 5, and 10 μg L^{-1}, respectively. In the study, the highest concentrations of these elements in leachates did not exceed 0.7, 0.07, and 0.20 μg L^{-1} for As, Cd, and Pb, respectively (Fig. 7.9). This fact confirms that applying sludge that meets the current criteria for use in agriculture creates no risk of groundwater contamination with potentially toxic trace elements, even when using reclamation sludge rates.

Summarizing, the long-term lysimeter study revealed that no sewage sludge toxicity was observed for plants, irrespective of the sludge rate, and straw and grain yield, in general, substantially increased after sludge application to the levels

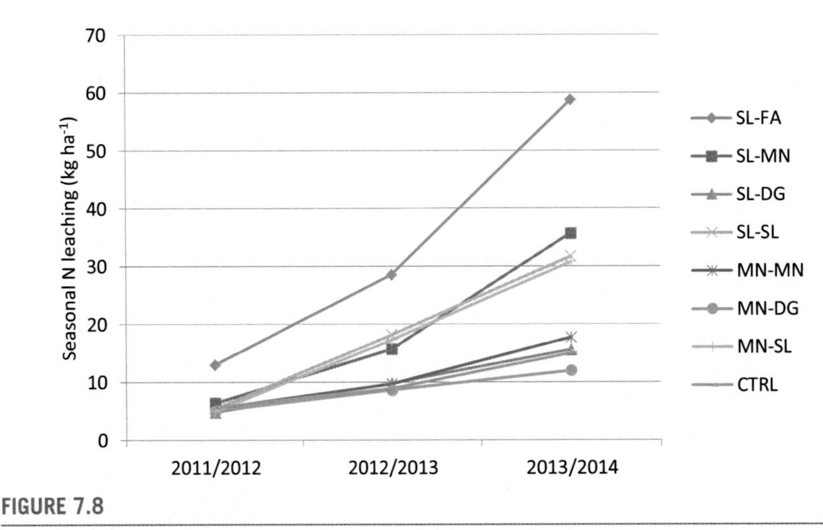

FIGURE 7.8

Leaching of nitrogen as dependent on fertilization variant.

equivalent to mineral fertilized fields. The concentration of nitrates in leachates and the total leaching of nitrogen from soil fertilized with high, reclamation sludge rates was slightly higher than in the case of mineral fertilization. Concentrations of trace elements in leachates were many times lower than those allowed for drinking water. The use of high reclamation sludge rates promoted permanent carbon sequestration in the soil. Sludge, applied even in reclamation rates, stimulates the activity and abundance of soil with microorganisms, especially bacteria.

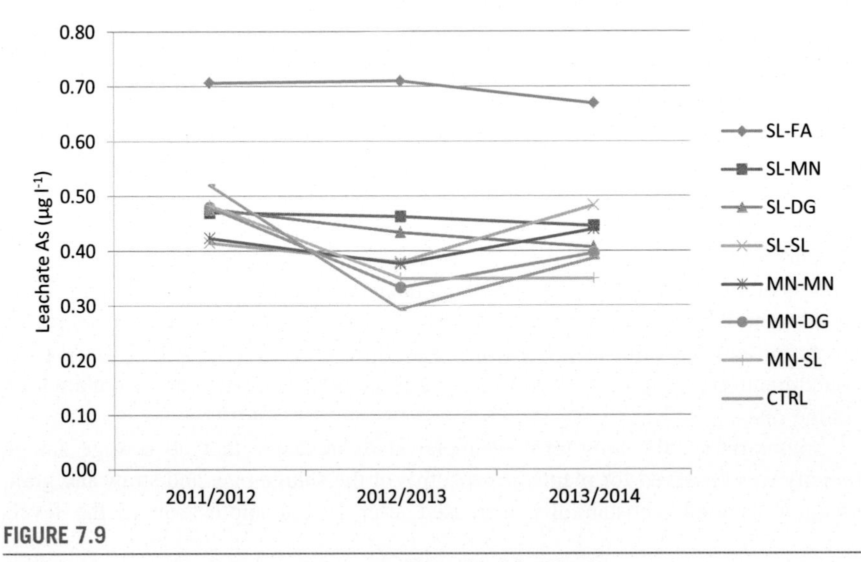

FIGURE 7.9

Arsenic concentration in leachates as an effect of soil fertilization.

Conclusions

Long-term experiments located in Poland enable long-term observation of impacts of crop production systems, rotation, or EOM application on soil quality in the context of Polish soil and climatic conditions. The history of some experiments dates back to 60 years. They provide rich data on impact of rotation and tillage on soils and crop performance. The experiments revealed that the simplification of soil tillage leads to a decrease in yields, but the response varied across crops. The highest yield decreases were observed in the initial period of tillage simplification, while in the subsequent years, the yield stabilization was observed. Simplifications in tillage generally increased weed abundance. On the other hand, the reduced tillage improved soil organic matter content and soil microbiological activity.

Other studies revealed that organic system of production did not ensure substantial soil organic matter increase. Depletion of soil available phosphorus and potassium was observed in the organic system, but on the other hand, soil physical parameters and soil microbial activity were improved as compared with conventional systems.

Regional long-term assessments, surprisingly, showed SOC accumulation in most of locations distributed across arable land in Dolnoslaskie region, especially in soil with initial low SOC content. The increase was observed despite the decrease of manure rates related to abandonment of animal production in the region. It can be explained by the substantial rise of crop yields and the related increase in amount of crop residues left in soil.

The studies also documented that application of good quality sewage sludge does not pose a risk of excessive trace metals leaching to groundwater quality even if reclamation rates are applied. Sewage sludge application stimulated soil microbial activity and soil carbon sequestration. However, attention must be paid to the leaching of nitrates.

References

Bień, J.D., 2012. Management of municipal sewage sludge by thermal methods. Inz'ynieria i Ochrona Srodowiska 15, 439–449 (in Polish).

Blecharczyk, A., Małecka-Jankowiak, I., Sawinska, Z., Piechota, T., Waniorek, W., 2018. A 60-year fertilization experiment on plant cultivation in crop rotation and continuous croping system in Brody. In: Marks, M., Jastrzebska, M., Kostrzewska, M.K. (Eds.), Long-term Experiments in Agricultural Studies in Poland. UWM Publications, pp. 27–40 (in Polish).

GUS, 2018. http://www.stat.gov.pl/bdl/app/strona.html?p_name=indeks. (Accessed 10 January 2018).

Kaczyński, R., Siebielec, G., Hanegraaf, M.C., Korevaar, H., 2016. Modelling soil carbon trends for agriculture development scenarios at regional level. Geoderma 286, 104–115.

Król, A., Lipiec, J., Turski, M., Kuś, J., 2012. Effects of organic and conventional on physical properties of soil aggregates. International Agrophysics 27, 15–21.

Kuś, J., Jończyk, K., Stalenga, J., Feledyn-Szewczyk, B., Mróz, A., 2010. Yielding of selected winter wheat varieties in conventional and organic farming. Journal of Research and Applications in Agricultural Engineering 55, 219–223 (in Polish).

Kuś, J., Smagacz, J., 2018. Yielding of cereals in crop rotations with different cereals proportion in a long-term experiment in Grabów. In: Marks, M., Jastrzebska, M., Kostrzewska, M.K. (Eds.), Long-term Experiments in Agricultural Studies in Poland. UWM Publications, pp. 73–94 (in Polish).

Lenart, S., Radecki, A., Ciesielska, A., Perzanowska, A., 2018. The effect of long term zero tillage treatment on soil properties, weed infestation and crop yielding. In: Marks, M., Jastrzebska, M., Kostrzewska, M.K. (Eds.), Long-term Experiments in Agricultural Studies in Poland. UWM Publications, pp. 177–192 (in Polish).

Mackay, J.E., Cavagnaro, T.R., Jakobsen, I., Macdonald, L.M., Grønlund, M., Thomsen, T.P., Müller-Stover, D.S., 2017. Evaluation of phosphorus in thermally converted sewage sludge: P pools and availability to wheat. Plant and Soil 418, 307–317.

Małecka-Jankowiak, I., Blecharczyk, A., Piechota, T., Sawinska, Z., Waniorek, W., 2018. Long-term field experiments on ploughless cultivation in Brody. In: Marks, M., Jastrzebska, M., Kostrzewska, M.K. (Eds.), Long-term Experiments in Agricultural Studies in Poland. UWM Publications, pp. 193–206 (in Polish).

Marks, M., Jastrzebska, M., Kostrzewska, M.K. (Eds.), 2018a. Long-term Experiments in Agricultural Studies in Poland. UWM Publications, pp. 1–280 (in Polish).

Marks, M., Rychcik, B., Treder, K., Tyburski, J., 2018b. A50-year study on growing plants in crop rotation and continuous cropping systems – a source of knowledge and a monument of farmin culture. In: Marks, M., Jastrzebska, M., Kostrzewska, M.K. (Eds.), Long-term Experiments in Agricultural Studies in Poland. UWM Publications, pp. 41–56 (in Polish).

Martyniuk, S., Ksieżniak, A., Jończyk, K., Kuś, J., 2007. Microbiological characteristics of soil under winter wheat cultivated in an ecological and conventional system. Journal of Research and Applications in Agricultural Engineering 52, 113–116 (in Polish).

Niewiadomski, W., 1995. Science of crop rotation - the state and prospects. Postepow Nauk Roln 3, 127–138 (in Polish).

Sadowski, C., Lenc, L., Kuś, J., 2010. Fusarium ear blight and fungi of the genus Fusarium colonizing the grain of winter wheat, a mixture of varieties and spelled wheat cultivated in the ecological system. Journal of Research and Applications in Agricultural Engineering 55, 79–84 (in Polish).

Siebielec, G., Siebielec, S., Lipski, D., 2018. Long-term impact of sewage sludge, digestate and mineral fertilizers on plant yield and soil biological activity. Journal of Cleaner Production 187, 372–379.

Siebielec, G., Stuczynski, T., 2008. Trace metal aspects of biosolids use. In: CD Proceedings of Protection and Restoration of the Environment IX. Kefalonia, Greece, ISBN 978-960-530-104-0, p. 7.

Stalenga, J., 2007. Applicability of different indices to evaluate nutrient status of winter wheat in the organic system. Journal of Plant Nutrition 30, 351–365.

Stepień, W., Sosulski, T., Szara, E., 2018. Interaction between mineral and organic fertilisation in long term experiments in Skierniewice. In: Marks, M., Jastrzebska, M., Kostrzewska, M.K. (Eds.), Long-term Experiments in Agricultural Studies in Poland. UWM Publications, pp. 11–25 (in Polish).

VDLUFA, 2014. Humusbilanzierung. Eine Methode zur Analyse und Bewertung der Humusversorgung von Ackerland (Standpunkt). VDLUFA, Speyer.

CHAPTER

8

Geographical network: legacy of the Soviet era long-term field experiments in Russian agriculture

Vladimir A. Romanenkov[1,2,3,6], Lyudmila K. Shevtsova[2,4], Olga V. Rukhovich[2,5], Maya V. Belichenko[2,6]

[1]*Lomonosov Moscow State University, Soil Science Faculty, Eurasian Center for Food Security, Moscow, Russia;* [2]*All-Russian Institute of Agrochemistry named after D. Pryanishnikov, Geographic Network of Field Experiments Department, Moscow, Russia;* [3]*Head of Department, Agrochemistry and Plant Bochemistry, Lomonosov Moscow State University, Moscow, Russia;* [4]*Principal Researcher, Geographic Network of Field Experiments Department, All-Russian Institute of Agrochemistry named after D. Pryanishnikov, Moscow, Russia;* [5]*Head of Department, Geographic Network of Field Experiments, All-Russian Institute of Agrochemistry named after D. Pryanishnikov, Moscow, Russia;* [6]*Leading Researcher, Geographical Network of Field Experiments, All-Russian Institution of Agrochemistry named after D. Pryanishnikov, Moscow, Russia*

Introduction

The need to study agrocenosis as a system with interacting components to identify functional relationships and find the ways to manage them led to the appearance of long-term field experiments more than 175 years ago. These experiments were agronomically oriented and intended primarily for the development of appropriate methods of crop cultivation under particular natural conditions. Initially, they included a limited list of observations of plants and soils. However, they laid the methodological foundation for assessing agroecosystems and their sustainability under the influence of agronomic impacts over the years and decades (Rasmussen et al., 1998). At present, the need—in ongoing as well as new experiments—of a more detailed characterization of climatic and soil conditions is obvious (Körschens, 2006; Richter et al., 2007; Johnston and Poulton, 2018). It is important to assess the efficiency of different elements of intensive technologies implemented in agriculture. It is also important to expand research on hayfields, pastures, perennial plantations, and in natural areas, as well as to gain maximum information from the ongoing experiments to solve both fundamental and applied problems (Janzen, 2009).

Long-Term Farming Systems Research. https://doi.org/10.1016/B978-0-12-818186-7.00009-6
Copyright © 2020 Elsevier Inc. All rights reserved.

For the most efficient use of this information, it is reasonable to organize a unified experimental research network for comparative assessments of crop response to the application of fertilizers, soil reclamation measures, crop rotation systems, and tillage technologies under different soil and climatic conditions. Such an experimental network makes it possible to determine the relationships between the components of agrocenoses in the plant–soil–climate system and to predict their response to various external disturbances, including soil degradation and progradation processes, global climate change, and interannual variations in heat and moisture supplies. This leads to timely and planned decisions rather than compulsory decisions on adaptation to such changes (Romanenkov, 2012).

These issues began to be intensively discussed in the international research arena since the 1990s (Powlson et al., 1996; Rasmussen et al., 1998). By that time, such a system in the Soviet Union had a 60-year-old history that had begun in the 1930s, when, on the initiative of D.N. Pryanishnikov and A.N. Lebedyantsev, methodological principles of the geographic network of experiments with fertilizers (Geonet) had been formulated (Pryanishnikov, 1943).

Inventory of current long-term experiments

The ongoing experiments with various fertilizer systems within the Geonet framework in different regions of Russia are shown in Fig. 8.1. At present, more than 130 long-term field experiments with fertilizers performed by 65 scientific institutions in Russia are carried out within the framework of the Geonet; more than half of them have at least a 35-year-long history; 10 of them last for more than 70 years.

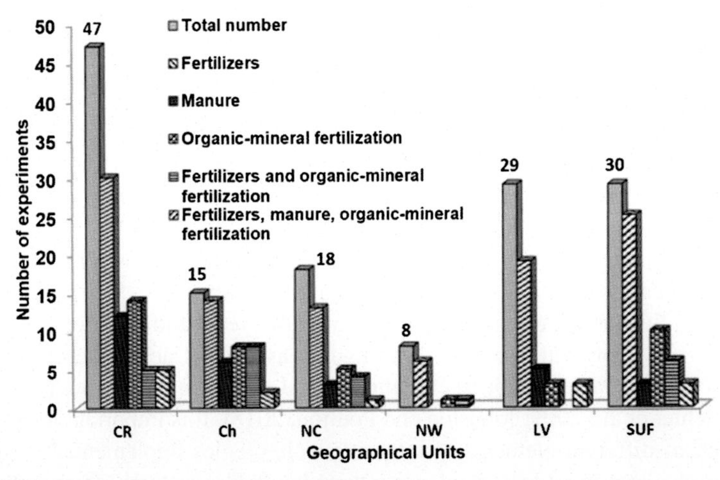

FIGURE 8.1

The number of long-term studies with mineral and organic fertilization systems undertaken across geographical units. *Ch*, Chernozem Region; *CR*, Central Russia; *LV*, Lower Volga; *NC*, North Caucasus; *NW*, North West; *SUF*, Siberia, Urals, Far East.

The greatest number of such experiments is being performed in the nonchernozemic zone of the Central Region and in the Volga Region. In the Siberian Region, there are 20 such experiments; in the Ural Region, 7 experiments; and in the Far East Region, 3 experiments.

Factors that are studied in long-term experiments (LTEs) in addition to various fertilizer systems are distributed differently depending on the region (Fig. 8.2). In the Central Russia, great attention is paid to liming and crop-protecting chemicals; in the Chernozem zone, to crop rotations and tillage systems; and in the North Caucasus, to tillage systems, management of crop residues, and irrigation. The experiments in the Volga Region more or less evenly consider all the factors; in most of them, the application of crop residues into the soil is specially studied, while one LTE is devoted to the effects of irrigation. In the Siberian Region, crop-protecting chemicals, tillage systems, and crop rotation systems are studied; one experiment is performed on the irrigated field. In the Ural Region, there are experiments with application of crop residues, green manuring, liming, and crop-protecting chemicals. In the Far East, the experiments with fertilizers also involve the study of the effects of liming, application of crop residues, and crop-protecting chemicals.

The Geonet as a unified experimental base for the study of fertilizers in Russia and, earlier, in the entire Soviet Union has a history of over 78 years. Although the historical order of the Peoples' Commissariat for Agriculture on the creation of the Geonet was issued in January 1941, it was preceded by a long path of the development of agricultural chemistry in Russia and abroad.

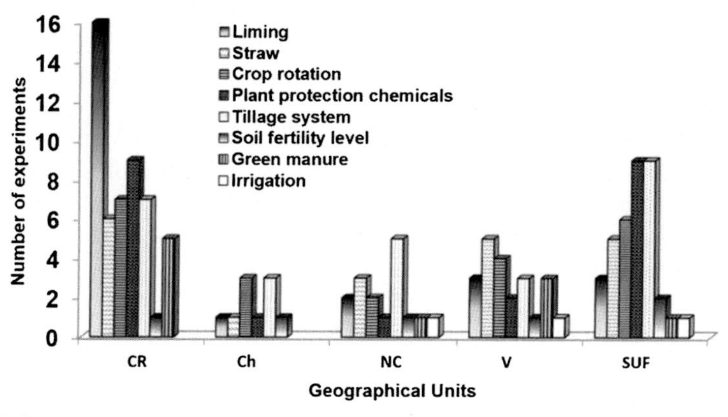

FIGURE 8.2

The number of long-term studies combining investigation fertilization effect with the other interventions. *Ch*, Chernozem Region; *CR*, Central Russia; *NC*, North Caucasus; *V*, Volga; *SUF*, Siberia, Urals, Far East.

Historical overview

In Russia, the history of field experimentation began at the end of the 19th century, when, on the initiative of local governments (zemstvos) and stakeholders, experimental fields belonging to these governments were organized. The first field trials appeared in 1884. Prominent scientists D.I. Mendeleev, K.A. Timiryazev, and A.G. Doyarenko played a significant role in the organization of field research in agronomy, agrochemistry, and plant physiology on these fields.

At the end of the 1890s, a permanent commission on agricultural experimental work was formed within the Ministry of Agriculture and State Property of the Russian Empire, which was charged with working out general principles for organizing official experimental institutions and subsidizing experimental institutions established by zemstvos. On the eve of World War I, there were more than a 100 experimental institutions employing nearly 1000 researchers in Russia.

In 1926–30, the Research Institute of Fertilizers (RIF) under the leadership of D.N. Pryanishnikov, A.N. Lebedyantsev, and A.P. Levitsky conducted 3808 short-term field experiments involving 317 experimental institutions according to the unified schemes and methods. The geographical network of experiments performed by the RIF served as a model of the cooperative work of agronomists, agricultural chemists, and soil scientists. The efficiency of fertilizers applied at the rates from 45 to 120 kg ha^{-1} on the main soil types was established; it was concluded that with a corresponding development of chemicalization in the north, in the podzolic zone, it is possible to obtain the same productivity as in the Chernozemic zone. The planning of the development of the fertilizer industry was based on the results obtained from this network.

Stationary experiments in Russia were planned on permanent experimental plots, considered as the basis for in-depth theoretical studies. In the 1920s, the main attention was paid to the adequate organization of experiments on the comparative study of the effect of manure and mineral fertilizers (Pryanishnikov, 1929). In the 1930s, it was shifted toward the study of the relationships between the dynamics of soil properties and the growth and development of plants with due account for the principle of compatibility of the results and their generalization. The creation of a geographic network of experiments with fertilizers required the development of uniform schemes and methods for conducting experiments and general principles for planning LTEs with fertilizers, as well as for planning soil, agrochemical, microbiological, and physiological studies on the experimental plots (Pryanishnikov, 1938).

With the further development of agriculture in Russia and the appearance of new challenges facing agricultural chemistry, the Geonet tasks have been enlarged: while maintaining the continuity of previously laid experiments, they have been further expanded from year to year by including multitude of factors addressing complexity of agricultural systems. This can be seen from the scheme in Fig. 8.3 and from a brief description of changes in the goals and results of research works conducted by the Geonet given below.

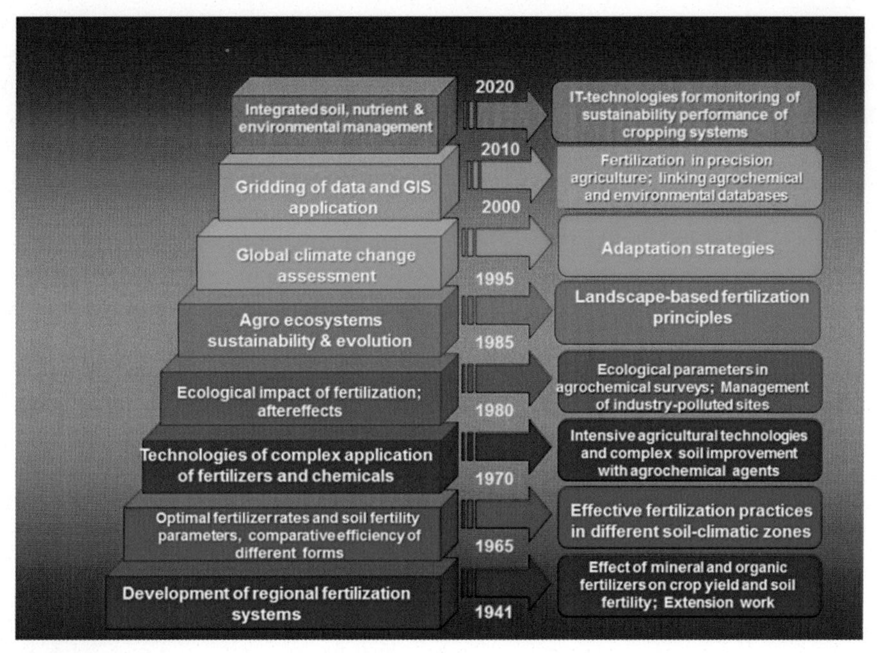

FIGURE 8.3

Main tasks and practical results of geographical network since foundation.

Source: Sychev et al. (2016), revised.

The organization of the Geonet at the initial stage envisaged the laying of field experiments according to a uniform methodology for zoning the rates of fertilizers required for the major crops, including the terms and methods for their application. At the same time, the efficiency of individual types of mineral fertilizers and their combinations was studied, and the effects of mineral and organic fertilizers were compared. The results were used to calculate the demand in fertilizers for different regions of the country and to plan the construction of new fertilizer plants. Fertilization systems for particular crops were developed, and first generalizations on the effect of fertilizers on plant development and soil fertility were made. The decisions of the choice of experimental plots were made on the basis of the materials of large-scale soil surveys of lands of collective farms performed in the 1930s–1940s; soil maps and agrochemical cartograms were used for this purpose.

The potential for increasing yields for different crops under optimal nutrition, crop rotations, and soil fertility level based on data generalization continues to be one of the current research fields. An assessment of the potential can be illustrated by comparing crop yield data on control plots of the Geonet (without application fertilizers for a long period), trials with the optimum rates of fertilizers (ensuring maximum yield), and statistical data on the yields of major crops (winter wheat, barley, and oats) for federal districts of the Russian Federation.

Thus, a comparison of data on the average yield of winter wheat in the experiments of the Geonet and in the farms (Fig. 8.4A) shows that in the Central and Chernozem Regions, the average yield in the farms is very close to the yield obtained on control plots of the Geonet. In the North Caucasus, South, and Volga Regions, the average yield in the farms is 0.5–1.3 t ha^{-1} lower than that on the control plots of the Geonet with the maximum difference in the South Region. LTEs with fertilizers performed at the Geonet demonstrate the potential crop yield in the particular large administrative units of Russia. In the North West Region, the real yield of winter wheat in the local farms is close to that on the optimally fertilized experimental plots, which attest to the high level of local farming ensured by the science-based rates of fertilization and appropriate crop rotation systems. In other regions, the differences between the average regional yields of winter wheat and the yields obtained in trials with optimum fertilization are quite significant and range from 2.9 to 5.6 $t \cdot ha^{-1}$.

For spring barley (Fig. 8.4B), in the North West, Central, and Siberian Regions, the average yields are higher than the yields on the control plots of the Geonet by 0.3–0.8 t ha^{-1}, which attests to a positive effect of fertilizers applied in the farms. In the North Caucasian, South, Volga, and Ural Regions, average yields of spring barley are lower than the yields on the control plots of the Geonet by 0.3–1.2 t ha^{-1}; the differences with the yield obtained on the optimally fertilized plots are higher and reach 1.0 t ha^{-1} in the Siberian Region and 2.6 t ha^{-1} in the South Region. The maximum difference between the average barley yield and the yield on the optimally fertilized plots is in the North Caucasian Region. This is explained by the application of the highly productive variety of winter barley in the experiments of the Geonet in this climatic zone. Significantly higher barley yields are obtained in the experiments in climatically favorable years.

For oats (Fig. 8.4C), the trends similar to those for spring barley are observed. However, as oats are less demanding for precursor crop than barley, the differences between the average yield and the yield on the control experimental plots are less pronounced. In the Chernozemic Region and in the Volga Region, these differences do not exceed 0.3 t ha; in the North West and Siberian Regions and in the nonchernozemic zone, they are insignificant. The yields obtained on the experimental plots with optimum fertilization are higher than the average yields in the districts by 0.6–2.4 t ha^{-1} (depending on the district).

The theoretical, organizational, and practical foundations were laid for the solution of large-scale tasks for the chemicalization of agriculture, which began in the Soviet Union in the 1960s. The chemicalization required a more detailed analysis of the efficiency of fertilizers for agricultural areas and different types of soil. It was important to find optimum rates of fertilizers for high-yielding varieties and for crops sown on fallow fields and after precursors (Naidin, 1968). Yield gains from the application of paired combinations of fertilizers had to be estimated. In general, these materials ensured possibility for assessing the effective soil fertility. The boundaries of fertilizer efficiency areas served as the basis for adjusting the boundaries of the soil-agrochemical areas. In the 1960s–70s, zonal fertilizer

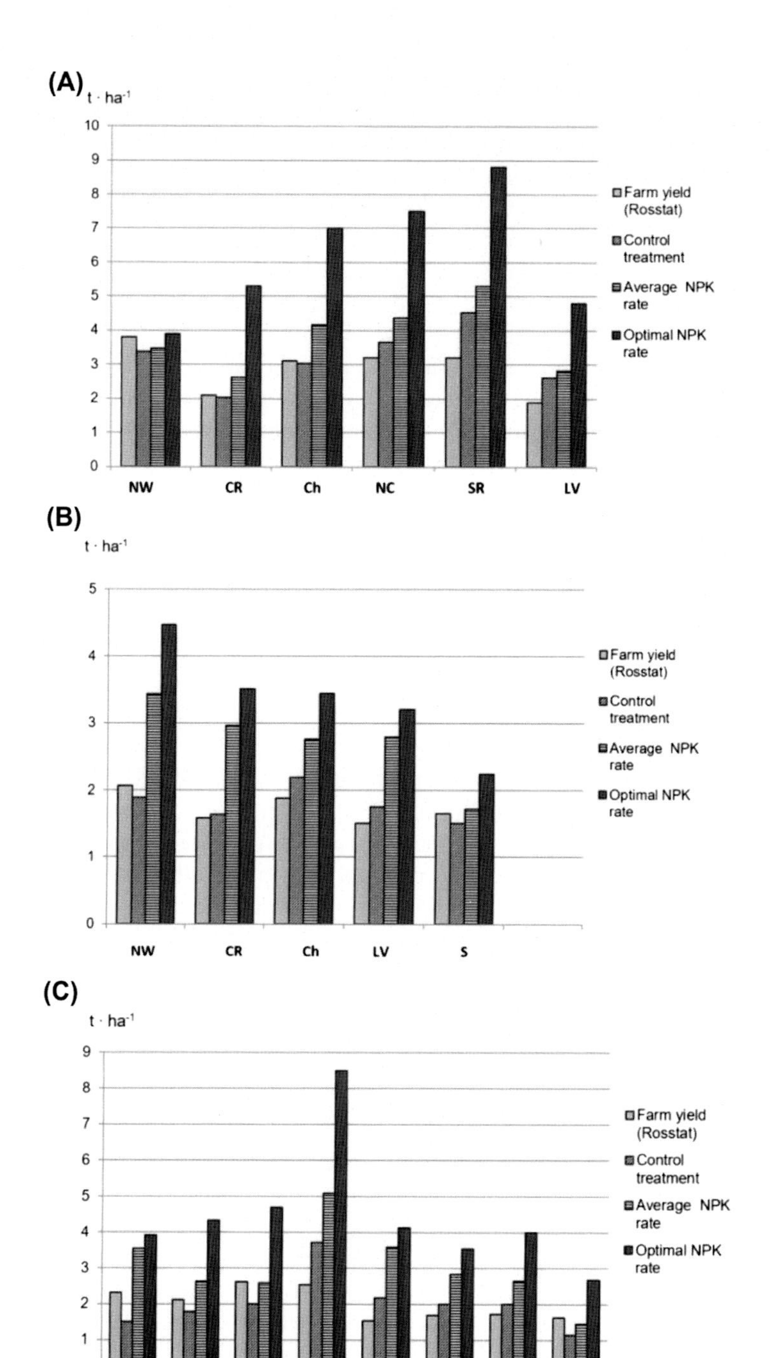

FIGURE 8.4

Comparison of winter wheat (A), oats (B), and barley (C) yield in geographical network LTEs with farm yields in 2005–15 across geographical units: *Ch*, Chernozem Region; *CR*, Central Russia; *LW*, Lower Volga; *NC*, North Caucasus; *NW*, North West; *S*, Siberia; *SR*, South Region; *U*, Urals.

systems were developed, and the main geographical patterns of fertilizer efficiency were determined with due account for the soil pools of nutrients and their mobilization capacity, the influence of climatic conditions, the duration of soil use and fertilizer application, and the applied agrotechnologies. In the 1960s, methodological recommendations of the Geonet for stationary field experiments were used in mass short-term experiments on the efficiency of different rates and combinations of fertilizers organized by the Agrochemical Service of the country. In turn, their results were taken into account in the soil-agrochemical zoning of the Soviet Union developed in the mid-1970s (Sokolov et al., 1966).

The development of intensive farming systems in the 1980s raised the question of assessing efficiency of the combined effect of fertilizers, different tillage technologies, land reclamation measures, and application of plant-protecting chemicals. These tasks required new methodological approaches, refinement of the ongoing and establishment of new field experiments. This problem was solved by the maximum geographic expansion of research at the experimental stations. This was the period when the Geonet involved the maximum number of participants. Overall, more than 320 experimental stations took part in the formulation and conduct of field experiments, including 159 stations in Russia, 54 stations in Ukraine, 28 stations in Kazakhstan, and 12 stations in Belarus. In total, more than 500 long-term agrochemical experiments were conducted in the former Soviet Union in those years. The results of field experiments performed by the Geonet served as the basis for the development of technologies aimed at improving soil fertility and its targeted management (Romanenkov and Shevtsova, 2014). In intensive farming systems, an important place was given to the factor of optimization of soil fertility parameters. In the soil—plant—fertilizer system, the factor of enhanced soil fertility management coupled with balanced plant nutrition with macro- and microelements, and compliance with environmental restrictions ensured high and stable yields of good quality.

The results of field experiments of the Geonet have widely been used as input parameters of regression models for calculating the rates of liming and fertilizers in dependence on the soil organic matter content and the contents of available phosphorus and potassium for maintaining the specified parameters of soil fertility in the intensive farming systems.

Optimization of soil fertility was achieved via application of moderate rates of fertilizers aimed to compensate for the removal of nutrients with the harvested crops. The rates of P and K fertilizers aimed at increasing the contents of available P and K in soils, i.e., at improving the soil fertility, were calculated with account for the soil texture, degree of acidity, cultivated crops and crop rotation systems, and applied agrotechnologies. In these calculations, it was taken into account that the transition of phosphorus into the unavailable nonexchangeable form is usually irreversible, whereas the ratio of nonexchangeable to exchangeable forms of potassium changes in time and space.

For the main soil types of the central region of the nonchernozemic zone, forest-steppe zone, central Chernozemic zone, and the North Caucasus, generalized data of the Geonet were used in regression models for different crop rotation systems ensuring productivity of up to 5–6 t (grain units)/ha for field crops and 8–9 t (grain units)·ha^{-1} for forage crops on moderately fertilized and fertile soils. This study took into account recommended rates of mineral fertilizers in intensive farming systems and the balance of soil nutrients (Romanenkov and Shevtsova, 2007).

An example of calculation of the optimal parameters of fertilizer systems for cereal–forb–row crop rotation systems is given in Table 8.1. A similar calculation for fodder crop rotations on soddy-podzolic soils indicated that the optimum productivity can be achieved at the rates of fertilization about 300–350 kg a.i. ha^{-1} per year with half of the major nutrients applied with manure. In crop rotations with 80% of row crops on moderately to highly fertile soils, the yield of fodder crops of 8.5 t ha^{-1} is achieved at these rates of fertilization with a payback of 16–21 kg fodder units per 1.0 kg NPK. From the economic point of view, it is preferable to apply moderate rates of fertilizers to the fertile soils than high rates of fertilizers to the low-fertile soils.

The results of LTEs on the integrated use of fertilizers and plant-protecting chemicals have been of great importance in substantiating the methods of increasing yields since the 1980s (Romanenkov and Shevtsova, 2014). Fig. 8.5 illustrates the possibility of controlling the yield and using fertilizer nutrients more efficiently in the case of appropriate plant protection system. In a year with favorable weather conditions, integrated and minimal plant protection systems of the Central Experimental Station of the All-Russia Research Institute of Agrochemistry (Moscow Region) ensured grain yields of 5.9–6.9 t ha^{-1} (Fig. 8.5A); the highest recovery of fertilizer N (60%–71% of the applied dose) was in the trial with the N application rate of 45 kg·ha^{-1} (Fig. 8.5B). The yield on the trials with plant-protecting systems increased by 40%–43%; both of the applied plant protection systems ensured approximately similar results. In dry years, the advantages of the integrated plant protection system were visible; it ensured the grain yield of 4.8–5.3 t ha^{-1} or 4.2–4.6 higher in comparison with the unprotected trials (Fig. 8.5C); the use of fertilizer N was 3–6 times higher in comparison with that in the trial with the minimal plant protection system (Fig. 8.5D).

In the 1980s–1990s, agroecological research topics became increasingly relevant. LTEs ensured a methodological possibility to assess an agroecosystem in terms of relationships between all its components (plant, soil, water, and air) and its stability under the influence of agronomic impacts over the years and decades. The results of the studies performed in those years were taken into account in the development of environmental indicators for the system of agroecological monitoring and in recommendations on the specificity of fertilizer systems on technologically polluted lands (Milashenko and Litvak, 1991).

This stage of the Geonet development was also marked by an expansion of experimental studies of the evolution of agroecosystems and adjacent environmental media based on a new type of experimental systems—agroecological polygons. The organization of such polygons made it possible to study the

Table 8.1 Optimal parameters of integrated nutrient management in crop rotation with cereals—grass—row crops providing sustainability of soil fertility and high yields in 80[s] 20th century, Geonet metadata.

Geographical unit, soil	Optimal average rate per year		Soil-agrochemical parameters			Nutrient balance, kg·ha^{-1} per year		
	Fertilizers kg·ha^{-1}	FYM, t·ha^{-1}	SOC (%)	P_2O_5 av.[a] (mg·kg^{-1})	K_2O av.[a] (mg·kg^{-1})	N	P_2O_5	K_2O
Central Russia, Retisols	180–250	10–15	0.9–1.5	100–150	120–170	5–10	80–90	0–5
Chernozem Region, Haplic Chernozems	150–200	5–10	3.5–4.6	100–150	80–120	−10–0	0–50	−25–−10
Luvic Chernozems	180–250	10–12	2.3–3.5	100–150	80–120	0–10	20–60	−20–0
North Caucasus, Haplic Chernozems	150–250	6–10	2.3–3.2	150–200	120–300	−5–0	20–60	−40–−20
Luvic Chernozems	220–280	9–12	2.0–2.3	150–200	120–180	−10–0	20–40	−30–−20

[a] Soil tests for available potassium and phosphorus in soils: the Kirsanov method (extraction with 0.2 N HCl) for acid soils of Central Russia; the Chirikov method (extraction with 0.5 M CH_3COOH) for soils of the forest-steppe zone; the Machigin method (extraction with 1% $(NH_4)_2CO_3$) for noncalcareous soils of the forest-steppe and steppe zones.

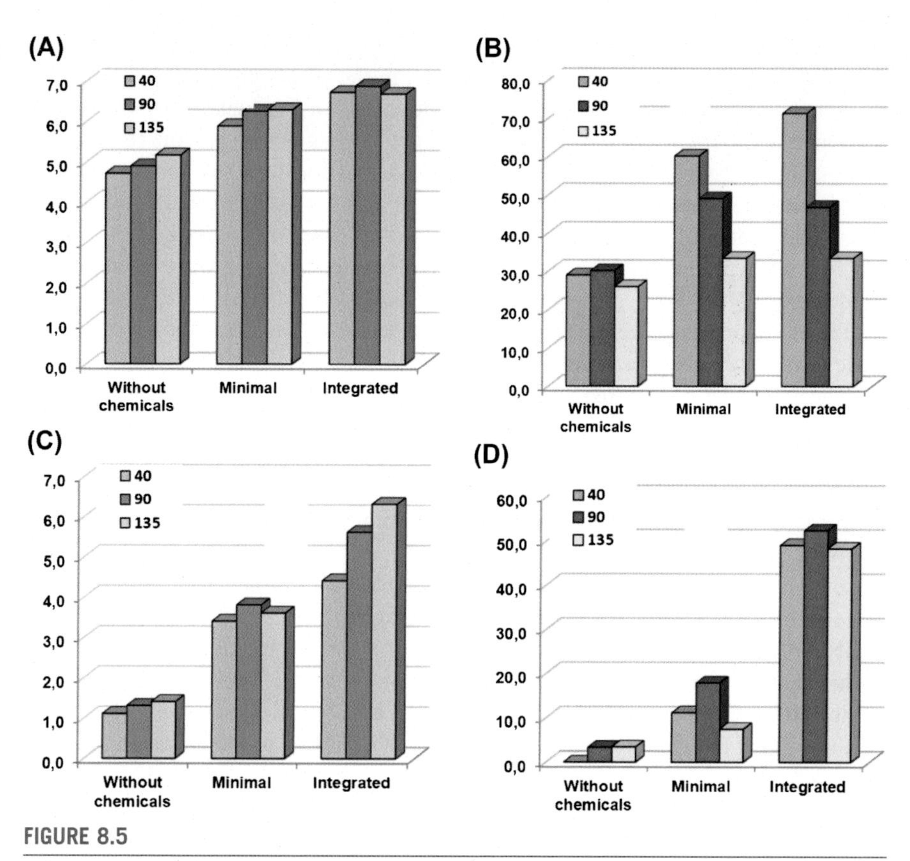

FIGURE 8.5

Effect of nitrogen fertilization (45–135 kg·ha^{-1}) and systems of chemicals application (minimal, integrated, control without chemicals) on winter wheat yield (a, c, t·ha^{-1}) and NUE (b,d, %) in favorable (A,B) and drought (C,D) years.

influence of the factors of agricultural intensification on the geochemically subordinate landscapes with due account for the geochemical flows of moisture, nutrients, and toxicants on slopes. The main purpose of the organization of agroecological polygons was their integration into the network of ground-based observations of the state of arable soils as an integral part of land monitoring. It envisaged a significant expansion of research on the agrophysical, biological, and phytosanitary conditions of the soil; the assessment of the intensity of erosional processes; and the monitoring of the development of secondary alkalinization and salinization. These studies were organized on 150 Geonet sites of agroecological monitoring. Their results laid the foundation for a landscape-adaptive approach in the application of fertilizer systems. The results obtained in the 1990s were taken into account in the methodological guidelines for integrated monitoring of agricultural land fertility, which are still in force (Sychev, 2006).

The extensive network of LTEs not only made it possible to study the relationships between the components of agrocenosis in the plant–soil system but also to predict their response to the interannual variability of heat and moisture supply. Stationary experiments of the Geonet turned out to be the most suitable for comparative studies of yield variability and weather conditions as they show long-term repeatability of the experimental patterns. Long-term experimental data of the Geonet and, later, of the Agrochemical Service were used by the Hydrometeorological Service to predict the efficiency of fertilizers under changing weather conditions (Fedoseev, 1979). With growing interest in the problem of climate change, the results of experiments were used in the development of the adaptation strategy for agriculture to the expected climate change.

The influence of weather conditions on the efficiency of fertilizers is largely determined by the level of soil fertility and the soil organic matter content as the most important source of nitrogen for plants. Statistical samples from the Geonet database prove the validity of this approach to estimate the efficiency of utilization of the agroclimatic parameters by the crops (Romanenkov et al., 2018).

With the help of such a database, spatial patterns of climatic risks and yield losses reflecting the agroclimatic features of agricultural production in the regions affected by droughts and the possibility of managing them by optimizing the conditions of plant nutrition and agricultural technologies were investigated. The database included information from 1993 to 2014 obtained from LTEs at 14 experimental stations of the Geonet in the Volga and North Caucasian Regions with the creation of a sample of the yield of grain crops arranged according to the increasing rates of fertilization.

The maximum absolute values of risk reduction because of optimization of plant nutrition at its initially high level were up to 25% for winter wheat, up to 15% for spring wheat, up to 11% for winter barley, and up to 6% for spring barley in the case of the moderate rates of N fertilization. At the same time, in most cases, the risk was reduced by one grade (for 12 temporal series) and by two grades for winter crops (for 4 temporal series) owing to a better realization of the bioclimatic potential by the crops (Table 8.2). This confirms the role of adherence to the proposed zonal technology of crop cultivation and crop rotation system, as well as the timeliness of operations performed as the measures of adaptation to drought conditions.

In this regard, in addition to research on the development of techniques that ensure sustainable production under adverse weather conditions, long-term stationary experiments provide an opportunity to evaluate agricultural soils as a source and sink of greenhouse gases and to search for possibilities to regulate matter and energy flows in agroecosystems via controlling agrogenic impacts (Shevtsova et al., 2015).

In 1940, special instructions for conducting field experiments at the stations included in the geographic network were issued. Later, they were periodically updated and published as *Guidelines for Conducting Research in Long-Term Experiments with Fertilizers*, in which the principles of site selection, its preparation, organization of the experiments, and the required minimum of field and analytical data were clearly stated. Since the establishment of the Geonet, scientific and methodological guidance and regional coordination of research have been carried out by the

Table 8.2 Spatial distribution of drought climatic risk of the yield of winter and spring wheat in 1994–2013 for Lower Volga and North Caucasus in percentage (%).

Winter wheat				
		Geonet climatic risks		
Region	**Farm climatic risks**	**Low N**	**Average N**	**High N**
Samara	20.7	15.4	12.5	12.5
Nizhny Novgorod	5.8	6.2	4.9	n.d.
Ulyanovsk	3.3	2.5	2.2	1.9
Saratov	27.5	1.3	2.1	0
Severnaya Osetiya	3.3	2.9	1.4	1.1
Stavropol Krai	12.3	4.3	5.9	2.5
Kabardino-Balkaria	5.5	5.1	2.8	2.5
Spring wheat				
Samara	26.0	24.1	20.1	19.3
Nizhny Novgorod	6.5	4.5	2.0	1.1
Ulyanovsk	3.8	3.5	3.0	3.0
Mordovia	9.2	6.4	2.8	n.d.
Saratov	36.7	27.3	22.0	22.1
Bashkortostan	9.8	9.9	8.3	n.d.

All-Union Institute of Fertilizers and Agropedology (at present, D.N. Pryanishnikov All-Russian Research Institute of Agrochemistry). This institute is responsible for the development of the Geonet, controlling its work, maintaining a permanently functioning information database of the experiments, and organizing efficient exchange of experience between the participants. On the basis of the inventory of LTEs performed in 1998–2000, a database of the passports of long-term field experiments has been created. The experiments with such passports are included in the "*Register of certificates of long-term experiments with fertilizers and other agrochemicals in the Russian Federation*." Currently, four volumes of this register are published in 2001–12. Long-term field experiments involve certain difficulties in their laying and further support, as it is necessary to ensure the required amount of field and analytical data and the continuity of research under replacement of their supervisors involving inevitable modifications of the experiments coupled with changes in the equipment and methodological approaches; often, the goals of further experiments represent a compromise with the original research goals.

Any changes in terms and conditions of funding, research priorities, and social conditions, as well as the lack of interested researchers are serious factors preventing the continuation of LTEs (Richter et al., 2007). Starting from the 1990s, Russian experiments have not escaped these problems. Many LTEs had to be closed or mothballed, including the oldest in Russia experiments initiated in the 1930s at the Dolgoprudnaya Experimental Station and Solikamsk Experimental Station.

On the initiative of scientists from European countries, the European network for LTEs on soil organic matter (SOM) has been created. It integrates information from 25 LTEs in Russia, 13 experiments in Ukraine, 3 experiments in Belarus, 5 experiments in Moldova, and 4 experiments in Georgia (Franko et al., 2002; Smith et al., 2002). Information from LTEs was integrated in the European SOM Network and LTSE Global Inventory systems and used in international and European projects for verification of dynamic models and predictive modeling, including better understanding and management of C cycles based on the dynamic modeling of soil C and crop growth for adaptation of agriculture to the expected climate change.

The static models of humus balance and their transformation which have been elaborated on the basis of LTEs in Russia for quantification known trends and work out new ones for significantly better understanding mechanisms of the agricultural factors' influence on soil organic carbon (SOC) (Shevtsova et al., 2003). It was used to estimate the influence of climate effects and changing agricultural practices on SOC levels in soddy-podzolic soils in the Russian Federation until 2050. It was revealed that among random climatic variables, affecting humus change, the most sensible ones are precipitation during vegetation period, potential evapotranspiration, and bioclimatic potential, which equals permanent grassland productivity under multicutting regime per vegetation period ($T > 5°C$). Fixed input effects include such management factors, as annual FYM rate for the modeling period, annual mineral N rate, percentage of row crops and fallow in the rotation structure, percentage of perennial grass in the rotation structure, and soil parameters—initial carbon content and the fraction of soil particles >0.01 mm. This work allowed defining carbon content levels for practical farmers and researchers as well as for advising and legislative agencies. The most important results are (i) a regular effect of climatic conditions on SOM dynamics in arable soddy-podzolic soil of European Russia and (ii) the possibility of simulation of optimal agricultural practices for sustainable C level for specific region. The regression-based model of humus balance was compared with dynamic model RothC in their description of steady-state conditions (Smith et al., 2007). With quite close estimates obtained, it was shown that both models demonstrate good agreement in assessing SOC equilibrium conditions over the next 50 years at the regional scale.

Future challenges

In the 21st century, with the development of computer-based mapping technologies and digital soil mapping, the elements of regular sampling grids are considered territorial information carriers; upon sufficient resolution of the grid, they allow the correction of agrotechnologies applied within a field. This technology includes georeferencing for linking with electronic maps and attributive databases, including remote sensing for forecasting yields and quality of crops within farms and individual fields. Pilot studies in this area prove the possibility of reducing the rates of mineral fertilizers without reducing the availability of nutrients because of minimization of the loss of fertilizers, more efficient use of their aftereffect, and maintenance of even crop growth. Adequate assessment of the spatial variability on the basis of research on individual fields allows one to develop fertilization strategies with due account for additional information on the state of the environment, topography, weather conditions, and landscape and soil characteristics (Romanenkov, 2012; Sychev et al., 2008). This implies further improvement of the principles of landscape-adaptive approach to the application of fertilizers and can be considered the major challenge of the fifth stage of the Geonet development.

An acute problem of the 21st century is the attainment of sustainable functioning of agrocenoses in space and time, which necessitates special studies on the integrated management of soil fertility. LTEs of the Geonet provide possibility to quantify the effects of the particular elements of agrotechnologies—mineral and organic fertilizers, ameliorants, crop protectors, tillage methods, and crop rotations—and to study long-term trends in crop yields, carbon sequestration capacity, efficiency of the use of applied nutrients, the functioning of soil biota, availability of microelements, and the structural and water-physical properties of soils. The practical use of the results obtained at the present time is that they allow us to substantiate the ecologically safe and economically feasible application of fertilizers and, thus, contribute to optimization of farming systems within individual farms, for example, based on feasibility assessment of best management practices.

In the future, the solution of these problems will be associated with the use of information technologies of monitoring systems. Field experiments can be considered an object of the local agroecological monitoring system, the management of which is linked to the tasks of the environmental protection and local pollution surveys (Sychev et al., 2011, 2016). The monitoring will be based on specially created geographic information systems that should include information on the state of the investigated areas at all the stages—from the laying of the experiment to its current state. The development of information technologies will require adequate linking of new high-resolution satellite images to the topographic base and to the existing schemes of landscape zoning. It will be important to link actual remote sensing data to the existing ground-based research points and to develop the principles of data extrapolation and interpolation. This information will be used in the systems of targeted management of soil fertility indicators and sustainability of agroecosystems. It is a challenge to expand the results of LTEs to a landscape level with the use of GIS tools and information technologies; in particular, it is important to delineate the most promising areas for agricultural intensification and target fertility management without damage to the environment.

The landscape-based approach should be used for a representative assessment and prediction of the state of the environment of the adjacent territories on the basis of the results of local field experiments.

For more than 60 years, the LTEs of the Geonet have served as the information base for the scientific support of the efficient, environmentally balanced, and economically feasible application of fertilizers, the importance of which significantly increases and goes beyond the scope of agricultural sciences because the evolution of agroecosystems is largely associated not only with the environmental and economic issues but also with the social conditions of the population (Table 8.3).

Table 8.3 Major directions of research in long-term experiments (LTEs) with fertilizers: past, present, and future.

Research area	Methodology	Issues to be solved	Limitations	Future challenges
Program of agrochemical monitoring	Unified system of indicators, sampling schemes, and analytical methods Integrated database Creation of statistical models	Zoning of crop growing systems Long-term trends of the quantity and quality of the yield Sustainability of soils and long-term changes in soil fertility Scientific basis for laying short-term experiments	Costly for continuation Inevitable changes in the experimental conditions Small size of the plots and their nearly level topography Marginal effects Insufficient assessment of spatial variability	Extrapolation and predictions in the agroecosystem monitoring Integration of data from long-term field experiments with regional information
Environmental issues in agrochemistry	Experimental long-term study of the plant–soil–groundwater–atmosphere relationships for the main agroecological zones Creation of statistical models Verification of dynamic models according to the long-term monitoring data	Long-term changes in the cycles, pools, and fractions of biogenic elements Impact of agrotechnologies on the slowly changeable soil properties Reproductions of the ecological soil functions New information on the basis of archive samples and databases Scientific basis for laying short-term experiments	LTEs were designed for testing the efficiency of fertilizers and soil fertility dynamics rather than for agroecosystem studies Limited set of agroecosystem parameters Changes beyond the experimental plot are not controlled Absence of data on the local parameters for model calibration	Interdisciplinary research and cooperation with specialists in neighboring sciences Spatiotemporal relationships from the level of microplot to the level of landscape Monitoring of greenhouse gases
Adverse changes in agroecosystems	Limits for various components of agroecosystems Assessment of the maximum anthropogenic load, including the use of wastes	Long-term trends in the accumulation, migration, and transformation of various pollutants (PAHs, heavy metals, oil products, etc.) Buffer properties of soils Criteria of agroecosystem degradation	Usually, local significance; difficulties in the choice of the optimum regional farming system Inconsistence of analytical methods Limited estimates of microbiological indicators	Discrimination between the background pollution and local pollution sources Study of the habitat of soil biota, gene pools, and the sources of biodiversity
Study of the impact of climate and weather conditions	Integration of long series of experimental field data and agrometeorological parameters Dynamic modeling on the basis of different time steps Integration of databases on meteorological, agronomic, and economic conditions of agriculture	Evaluation of the impact of climate change on sustainability of agroecosystems (yield trends, carbon pools, rates of carbon turnover, etc.) Creation of adaptation strategies for agriculture	Focus on the current state of resources rather than on the study of their evolution Absence of adequate approaches to assess the influence of particular weather conditions as a factor in the LTEs	Optimization of the distribution of experimental stations in time and space Identification of the most informative years in long-term temporal data series

References

Fedoseev, A.P., 1979. Агротехника и погода (Agronomical practices and weather). Gidrometeoizadat, Leningrad (in Russian).

Franko, U., Schramm, G., Rodionova, V., Körschens, M., Smith, P., Coleman, K., Romanenkov, V., Shevtsova, L., 2002. EuroSOMNET – a database for long-term experiments on soil organic matter in Europe. Computers and Electronics in Agric 33, 233–239.

Janzen, H.H., 2009. Long-term ecological sites: musings on the future, as seen (dimly) from the past. Global Change Biology 15, 2770–2778.

Johnston, A.E., Poulton, P.R., 2018. The importance of long-term experiments in agriculture: their management to ensure continued crop production and soil fertility; the Rothamsted experience. European Journal of Soil Science 69 (1), 113–125.

Körschens, M., 2006. The importance of long-term field experiments for soil science and environmental research: a review. Plant Soil and Environment 52, 1–8.

Milashenko, N.Z., Litvak, Sh.I., 1991. Методические и организационные основы проведения агроэкологического мониторинга в интенсивном земледелии (Methodological and organizational baselines of carrying agroecological monitoring in intensive agriculture). GKNT, Moscow (in Russian).

Milashenko, N.Z., Sychev, V.G. (Eds.), 2001–2012. Реестр аттестатов длительных опытов с удобрениями Географической сети опытов Российской Федерации (Register of certificates of long-term experiments with fertilizers and other agrochemicals in the Russian Federation), vols.1–4. VNIIA, Moscow (in Russian).

Naidin, P.G., 1968. Полевой опыт как метод изучения вопросов земледелия (Field experiment as a method for study agricultural issues). In: Field Experiment. Kolos, Moscow, pp. 5–27 (in Russian).

Powlson, D.S., Smith, P., Smith, J.U., 1996. Evaluation of soil organic matter models using existing, long-term datasets. In: NATO Advanced Research Workshop "Evaluation of Soil Organic Matter Models Using Existing Long-Term Datasets". Springer-Verlag, Berlin. Harpenden, England, NATO ASI series. Series I, Global environmental change. 38.

Pryanishnikov, D.N., 1929. О сравнении действия навоза и минеральных удобрений (About comparative effect of manure and mineral fertilizers). Udobreniye i urozhai 1, 8–15 (in Russian).

Pryanishnikov, D.N. (Ed.), 1938. Постановления совещания по вопросам организации стационарных опытов по изучению действия удобрений (Regulations of the Meeting on the Issue of Launching Long-Term Stationary Experiments to Study Effect of Fertilizers Application). VASKHNIIL, Moscow (in Russian).

Pryanishnikov, D.N., 1943. Столетие Ротамстеда – праздник агрохимической науки (The centenary jubilee of Rothamsted – a feast of agronomical science). Pochvovedenie 9–10, 77–92 (in Russian).

Rasmussen, P.E., Goulding, K.W.T., Brown, J.R., Grace, P.R., Janzen, H.H., Körschens, M., 1998. Long-term agroecosystem experiments: assessing agricultural sustainability and global change. Science 282, 893–896.

Richter Jr., D., Callaham, M.A., Powlson, D.S., Smith, P., 2007. Long-term soil experiments: keys to managing Earth's rapidly changing ecosystems. Soil Science Society of America Journal 71 (2), 266–279.

Romanenkov, V.A., 2012. Агрохимические опыты в системе исследований Геосети: прошлое, настоящее и будущее (Agrochemical trials in the Geonet research system: past, present and future). Izvestia Timiryazev Selskohoz Academy 3, 54–61 (in Russian).

Romanenkov, V.A., Pavlova, V.N., Belichenko, M.V., 2018. Оценка климатических рисков при возделывании зерновых культур на основе региональных данных и результатов длительных опытов Геосети (Climatic risks assessment based on regional data and results of long-term experiments of Geonet). Agrokhimia 1, 77–86 (in Russian).

Romanenkov, V.A., Shevtsova, L.K., 2007. Разработка научных основ эффективности систем удобрения и управления плодородием почв на основе длительных опытов Геосети (Development of scientific basis for efficiency fertilization systems and soil fertility management based on long-term experiments). In: Actual problems of agrochemical science. 75 anniversary of D.N. Pryanishnikov All-Russian Research Institute of Agrochemistry. VNIIA, Moscow, pp. 123–127 (in Russian).

Romanenkov, V.A., Shevtsova, L.K., 2014. Длительные опыты Геосети в современных и перспективных агрохимических и агроландшафтных исследованиях (Long-term Geonet Field Experiments in Modern and Projective Agrolandscape Research). Agrokhimia 11, 3–14 (in Russian).

Shevtsova, L.K., Romanenkov, V.A., Blagoveshchenskiy, G.V., Khaidukov, K.P., Kanzyvaa, S.O., 2015. Структура баланса углерода и биоэнергетическая оценка его компонентов в агроценозах длительных полевых опытов (Structure of carbon balance and bioenergetics assessment of it's components in agrocoenoses of long-term field experiments). Agrokhimia 12, 67–75 (in Russian).

Shevtsova, L., Romanenkov, V., Sirotenko, O., Kanzyvaa, S., 2003. Transformation study in arable soils based on long-term experiments in Russia: historical experience and international co-operation. Archives of Agronomy and Soil Science 49 (5), 485–502.

Smith, P., Falloon, P.D., Körschens, M., Shevtsova, L.K., Franko, U., Romanenkov, V., Coleman, K., Rodionova, V., Smith, J.U., Schramm, U., 2002. EuroSOMNET – a European database of long-term experiments on soil organic matter: the WWW metadatabase. The Journal of Agricultural Science 138, 123–134.

Smith, P., Smith, J.U., Franko, U., Kuka, K., Romanenkov, V.A., Shevtsova, L.K., Wattenbach, M., Gottschalk, P., Sirotenko, O.D., Rukhovich, D.I., Koroleva, P.V., Romanenko, I.A., Lisovoi, N.V., 2007. Changes in mineral soil organic carbon stocks in the croplands of European Russia and the Ukraine, 1990-2070; comparison of three models and implications for climate mitigation. Regional Environmental Change 7 (2), 105–119.

Sokolov, A.V., Rozov, N.N., Rudneva, Ye.N., 1966. Почвенно-агрохимическая карта СССР (USSR Soil-Agrochemical Map). Agrokhimia 1, 3–12 (in Russian).

Sychev, V.G. (Ed.), 2006. Система агроэкологического мониторинга земель сельскохозяйственного назначения (System of Agroecological Monitoring of Agricultural Lands). Russian Academy of Agricultural Sciences, Moscow (in Russian).

Sychev, V.G., Rukhovich, O.V., Romanenkov, V.A., Belichenko, M.V., Listova, M.P., 2008. Опыт создания единой систематизированной базы данных полевых опытов агрохимслужбы и Геосети « Агрогеос » (The problem of formation unified and structured database "Agrogeos" of agrochemical service and geographic network of field experiments). Problems of Agrochemistry Ecol 3, 35–38 (in Russian).

Sychev, V.G., Yefremov, E.N., Romanenkov, V.A., 2011. Состояние и перспективы мониторинга земель сельскохозяйственного назначения и рационального использования потенциала почвенного плодородия (Monitoring of agricultural lands and rational use of soil fertility potential: current status and future prospects). Problems of Agrochemistry Ecology 4, 42–46 (in Russian).

Sychev, V.G., Yefremov, E.N., Romanenkov, V.A., 2016. Monitoring of soil fertility (agroecological monitoring)). In: Mueller, L., Sheudshen, A., Eulenstein, F. (Eds.), Novel Methods for Monitoring and Managing Land and Water Resources in Siberia, Springer Water. Springer, Switzerland, pp. 541–561.

CHAPTER

Managing long-term experiment data: a repository for soil and agricultural research

9

Meike Grosse, Carsten Hoffmann, Xenia Specka, Nikolai Svoboda
Leibniz Centre for Agricultural Landscape Research (ZALF), Müncheberg, Germany

Introduction

Agricultural long-term experiments (LTEs) are essential research infrastructures to reveal effects of agricultural management on soil processes in the long run. In many cases, long time series of research data exist, which also enable to model slowly occurring changes in the soil. LTEs provide important information for soil functions and services, such as the storage and filtering of carbon, nutrients and water, the habitat function, and productivity. LTEs enable to reveal effects of climate change on agricultural systems and are, not least, important for future food security. The joint analysis of data from several LTEs may lead to even more useful results. However, information about existing LTEs is scattered and the data are often not easy to access.

Measures to ensure that LTEs can be found easily and the provision of LTE research data in a central database are two aims of the research project "BonaRes." BonaRes is short for "Soil as a Sustainable Resource for the Bioeconomy." It is funded by the German Federal Ministry of Education and Research (BMBF). The resource efficient and responsible use of the soil is in the focus (BMBF and BMEL, 2014). BonaRes consists of 10 interdisciplinary research project consortia and the "BonaRes – Centre for Soil Research." Since 2015, the 10 project consortia have been developing new approaches to questions of sustainable soil use. A significant part of the research is based on data produced by LTEs. The BonaRes Centre for Soil Research aims to substantially improve the opportunities to exchange scientific knowledge and is the coordinating project (BMBF, 2018). An internet platform is established by the BonaRes Centre: the BonaRes Portal.[1] It has different subportals which provide, e.g., information about the BonaRes projects, access to knowledge and models, to decision support options for a sustainable soil management, and access to research data including data from LTEs.

[1] https://www.bonares.de.

Long-Term Farming Systems Research. https://doi.org/10.1016/B978-0-12-818186-7.00010-2
Copyright © 2020 Elsevier Inc. All rights reserved.

The BonaRes Data Centre,[2] a part of the BonaRes Centre, established a Data Repository for soil and agricultural research data. The BonaRes Repository focuses on the integration of research data from the BonaRes research program as well as from other researchers within this research area. Beyond that, special attention is drawn to the provision of information about research data from LTEs in Germany via the BonaRes Repository. Although special effort is made for the integration of data from German LTEs, the repository is open to research data from around the world.

Most of the times, data from LTEs are not described consistently, are not accessible for third parties, are structured by different formats and systems, and often exist only in analog formats, e.g., in field books (Table 9.1). It is, therefore, a great challenge to compile this scientific heritage in accessible and standardized formats for scientific and socioeconomic reuse with low barriers. According to the state of the art in international research data management and integration of data in an interdisciplinary context, LTE data should be findable for the research community, easily accessible by download functions, interoperable with other data and repositories, and reusable, e.g., for modelers. These four requirements are known and described as FAIR data principles in data science (Wilkinson et al., 2016). Standardized metadata, free and established vocabularies, standardized communication protocols, and, not least, a clear and accessible data usage license are provided by the BonaRes Repository.

Overview of long-term experiments in Germany

As there was no comprehensive information about LTEs and respective data in Germany, the aim was to compile such metainformation and make it available at one location.

Table 9.1 Current and potential characteristics of long-term experiment data in Germany.

	Past	Future: FAIR data
Metadata	Sporadic	Complete
Data localization	Decentralized	Central data repository
Data visibility	Low	High
Data accessibility	Difficult	Easy
Data citeability and reusability	Difficult, diverse	Easy

[2] https://datenzentrum.bonares.de/research-data.php.

The definition of LTE in the context of BonaRes is a minimum duration of 20 years and a static design. LTEs which are at the moment below the age of 20 years but have a planned duration of at least 20 years are also included. Attention is given exclusively to LTEs in the context of soil research and bioeconomy, i.e., objects of research should be as a minimum the soil properties and processes and the yield. The research themes of the LTEs were assigned to fertilization, tillage, crop rotation, "other" themes, and combinations of these according to the experimental factors. The setup of each trial should allow statistical analysis, i.e., have factors, treatments in a randomized design, and replications. Sources for the research were scientific paper as well as other articles, books, trial guides, and websites.

For the coordination and simplification of the metadata acquisition, the BonaRes Fact Sheet was developed, which asks for all relevant trial information such as setup, sampling, and site information (Grosse et al., 2019). Up to January 2019, 40 fact sheets were completed by different trial owners. The information in the fact sheets was processed within the BonaRes Data Centre.

An overview of the collected information is presented in the following section.

In total, 200 LTEs across Germany according to the definition above were identified (status December 05, 2018) (Table 9.2). The three oldest LTEs with a duration

Table 9.2 Overview of long-term experiment in Germany (n = 200).

Trial status	Ongoing	141
	Finished	59
Duration	>100 years	3
	50–99 years	53
	20–49 years	114
	<20 years	12
	Unknown	18
Land use	Field crops	163
	Grassland	34
	Vegetables/pomiculture	3
Research theme[a]	Fertilization	154
	Tillage	37
	Crop rotation	31
	Other	23
Holding institution	University/University of applied sciences	95
	State authorities	62
	Non-university scientific institution	22
	Industry	21
Farming category	Conventional	179
	Organic	14
	Integrated	5
	Conventional and organic	2

[a] *Multiple mentions possible (n = 245).*

of more than 100 years are "Ewiger Roggen" Halle (1878—today), the "Static Fertilizer Experiment Bad Lauchstaedt" (1902—today), and the "Long-term Fertilizer Experiment Dikopshof" (1904—2009). Twelve LTEs have not reached the age of 20 yet, but have a planned duration of 20 years or more. The duration of 18 finished trials is unknown as only the start of the trial is known but not the exact year of the end (Table 10.2). As these trials are mentioned in different important sources as ongoing (Amberger and Gutser, 1976; Körschens 1990, 1994, 1997, 2000; Debreczeni and Körschens, 2003), it is known that they were running at least 20 years.

The type of land use in most LTEs is arable field crops. Especially long-term grassland experiments are often dedicated to other scientific issues than soil and yield and are not included in this research then.

Most of the LTEs were established for research in fertilization. Historically, questions of the effects of fertilization on plant growth were in the focus of research, while more recently also the effects of crop management on soil properties were investigated (Herbst and Merbach, 2015). Most of the tillage experiments compare different tillage intensities. Often, the tillage experiments are younger experiments as 18 tillage experiments started in 1990 or later. However, the oldest tillage experiment started in 1965.

The other research themes than fertilization, tillage, or crop rotation are rather diverse. "Environmentally friendly plant protection" is the most frequent research theme among them (n = 5). "Irrigation" is the second frequent (n = 4).

More details of the situation of LTEs in Germany are presented in Table 9.2.

Information from the literature research and the Fact Sheets were compiled in a dynamic GIS-based web application, the overview map for LTEs, which is part of the BonaRes Repository.[3] The map offers different search and display options. The map contents can be displayed according to their research theme, duration, type of land use, farming category, or participation in networks. A selection of a single LTE gives general information about the trial. If available, the information from the fact sheets is displayed here. Besides the exact position of the LTE, the following information is shown: trial site, trial name, start of the trial and possibly end, trial holder, and research parameters.

The compilation and publication of this information in the BonaRes Repository shall improve the visibility of the LTEs and lead to a better reusability. This may make the LTEs themselves more valuable and may also help to maintain the LTEs.

Research data provision

Besides an overview with information on all LTEs in Germany, the acquisition of research data from single LTEs (in and outside Germany) is a task of the BonaRes Data Centre. The proceeding with standards, data policy, and the

[3] https://ltfe-map.bonares.de.

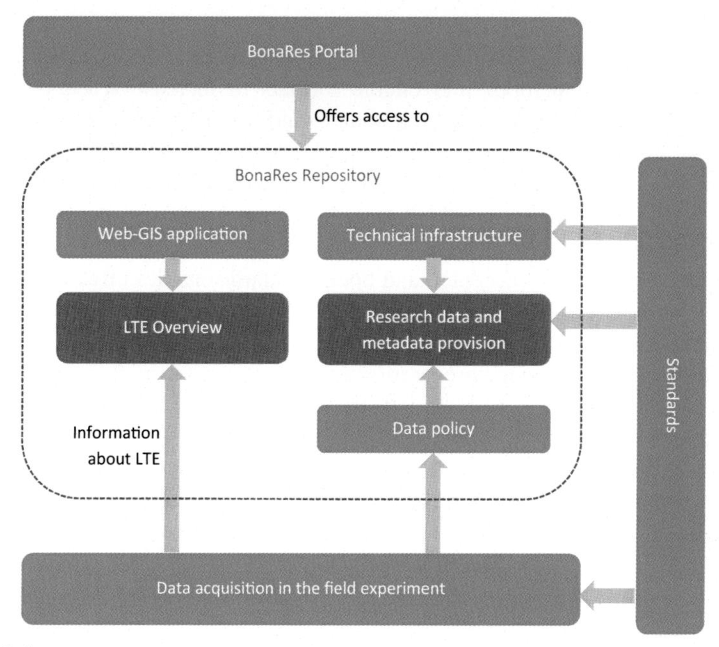

FIGURE 9.1

Schematic representation of the BonaRes Repository. The basis is the data acquisition in the field. Information about the field experiments are compiled and presented via a Web-GIS application. After signing a data policy, research data and metadata can be published for free reuse. Standards are applied at all data life stages and at the repository itself. The BonaRes Portal offers access to the repository.

technical implementation is described in the following section. For a schematic overview of the BonaRes Repository, see Fig. 9.1.

Overview of standards

To meet the FAIR data principles, it is necessary to apply free and widely used standards for all stages of LTE data life, from its acquisition in the field, via its structuring and transfer to the BonaRes Repository, until its provision via the BonaRes Data Portal. Besides this convention and requirements of using standards and designating them in the metadata, there are many advantages for providers and users of standardized research data: Standards facilitate cross-disciplinary data exchange, increase data comparability and visibility, enhance resource efficiencies of data storage, and enable interoperability with data of other national and international infrastructures.

The BonaRes Repository provides user-friendly upload and download functionalities for both data owner and data users. Data owners should provide detailed metadata including used methods and citable formats to increase reusability. As an agreement for all disciplines, research data should be collected under standardized conditions, stored with quality checks, described by coherent metadata, and be open for any scientific reuse (Alliance of German Science Organisations, 2008; Berlin Declaration Open Access, 2003). The compliance of these agreements ensures safe and long-term data reusability for processing, exchange, and review. The use of internationally accepted and open standards enables effective data management and the accessibility for different user groups in both typical fields of later data application, public and science.

In particular, the enhanced preparation and provision of research data from LTEs for reuse is very attractive, as LTE data are of such high value. There is a growing demand for open, accessible, and well-described data from LTEs. The scientific community is interested in comprehensively described and accessible data, e.g., to achieve best possible outcomes in soil productivity models, to assess the role of soils for yields under climate change processes, or to highlight habitat or filter function of soils. However, soil and agricultural research data, which were collected within LTEs, are mainly stored decentralized in heterogeneous formats and are hardly findable and accessible from outside.

To overcome this data problem, the BonaRes Data Centre supports LTE data owner during the data standardization and documentation process, e.g., data structuring and storage, metadata creation, data provision, data publication, and long-term archiving. The Data Centre recommends collecting research data under standardized conditions, to describe them with standardized metadata, to proof them by normalized quality tests, and to store them in the BonaRes database, to be visible and accessible for any data reuse (e.g., modeling), exchange, and further review.

The BonaRes Data Centre compiled, discussed, and published a list of more than 600 standards, regulations, code lists, and controlled vocabularies in the field of soil, agricultural, and data science to a report (Hoffmann et al., 2018) including recommendations for their application within the BonaRes project. Examples of relevant standards are given in Table 9.3.

Data life stages

Data acquisition

The primary data generation (acquisition) of soil-, crop-, yield-, fertilizing-, and machinery data from LTEs strongly depends on the research objectives and laboratory facilities. Soil and agricultural field and laboratory methods are widely regulated by numerous standards, laws, guidelines, and conventions. Some standards and regulations are used only in national contexts, such as soil classifications and plant variety names, and may require transformation to be comparable in international contexts. However, in terms of the LTE running times of often several decades, it is always

Table 9.3 Examples of relevant standards for the different data life stages.

Data Acquisition	- *Soil classification:* Mostly developed nationally: Germany (Survey Guideline KA5 (Ad-hoc-Arbeitsgruppe Boden (2005)), international: WRB: World Reference Base for Soil Resources (FAO and IUSS, 2015) - *Soil sampling*: Official standards, guide- or method books, e.g., ISO 18400-100:2017 - *Agricultural soil management:* Operation Planning Agriculture 2014/15 (KTBL, 2014), Good Agricultural Practices GAP (FAO, 2003)
Data Management	- *Exchange languages:* Based on XML and UML, they are widely used and accepted. Examples are AgroXML and SoilML (ISO 28258:2013, in revision). - *Validation and data quality*: e.g., statistical outliers, plausibility and consistency tests (e.g., ISO 16269-4:2010), geodata quality (ISO, 19157:2013) - *Transformation tools:* They may help to integrate national data into international systems
Data Provision	- *Geodata services:* Long-term experiment data with spatial information should be provided by free, widely distributed, and well-established OGC services (WMS, WFS, CSW) - *Access and reusability:* Open and standardized metadata schema based on INSPIRE and DataCite (see metadata) - *Controlled vocabularies:* AGROVOC (FAO), GEMET (EU), QUDT (NASA)

important to also identify used standards in the past and clearly allocate recently used methods in the metadata. All additional information to methods is valuable for later data reuse, e.g., in which year a standard or method changed or a new laboratory device or field instruments was used.

Data management

For assignable data management, standard procedures in quality control, database structure, data formats, and data transfer languages are strongly recommended. Data storage is generally possible in any format and structure, as long as usability and interpretability for future reuse is assured. However, the use of a standardized format and structure for data upload enables comparability of data from different LTEs. To meet this advantage, a data structure for LTE data was prepared by the Data Centre, which is provided to the data owner. It enables easy quality checks, clear data interpretation, consistent use of standardized codes, and data relations via primary and foreign keys. To avoid transfer errors to the database and to assure interoperability of LTE data with other databases, internationally accepted codes are used for several parameters. Standardized coding systems enable unique identifications of certain features such as machinery (Association for Technology and Structure in Agriculture, KTBL), plant varieties in Germany (the Federal Plant Variety Office [BSA] and the International Union for the Protection of New Varieties of Plants [UPOV]), or crops (the International Crop Codes ICC by the FAO).

Data provision

After standardized data acquisition and the transfer to the Data Centre, the supply of standardized metadata is necessary to clearly describe LTE datasets. Keywords should be provided according to controlled vocabularies. Thesauri with semantic relations improve data interlinkage, increase data visibility, and facilitate data exchange, networking, and communication between different repositories, also at international level. LTE data are always connected with spatial information (referred to as "geodata") and should be integrated into a geoportal with geodata services for user upload and download facilities. The Open Geospatial Consortium (OGC) provides numerous open and widely used standards. Before publishing data, notices on data license are essential for user-friendly data provision.

Data policy for long-term experiments

In accordance with its philosophy of providing open access to research data, the BonaRes Repository steadfastly considers that the data collected in LTEs should be made freely available for the best possible facilitation of scientific progress, so that researchers can reuse them wherever possible.

The free availability of research data enables validation, replication, reanalysis, new analysis, reinterpretation, or inclusion in metaanalyses and also facilitates the reproducibility of research. Providing data for these purposes enhances the value of these data from LTEs, much of which is funded from public sources. Ultimately, the position of the BonaRes Repository is quite simple: ensuring access to existing research data should be an essential part of the scientific work process.

The generic data policy for LTEs of the BonaRes Repository specifies the rules of "good scientific practice" related to data management as described by the German Research Foundation (DFG) and the German Research Alliance. These rules are implemented by the BonaRes Centre and the BonaRes Repository (Svoboda and Heinrich, 2017). The generic data policy specified for LTEs is available on request.

This data policy determines the transfer of data, creates the legal basis for public availability, and secures the right of first use for the owner of research data from LTEs. In addition, it spans out a framework within which the specific needs of data owner can be addressed.

Before data are transferred to the BonaRes Repository, the data policy must be approved by the respective institution's head on behalf of the responsible institution. This ensures that the data policy remains valid even in the event of a change of personnel, as the responsibility remains principally vested in the institution.

By agreeing to the data policy, the data owners accept that they or their representatives are responsible for the quality of the research data provided. The BonaRes Repository is very interested in high-quality data. The data stored in the repository must have passed through multiple quality assurance processes.

All high-quality research data to be stored in the BonaRes Repository must be described with metadata. The BonaRes Metadata Schema is mandatory (Gärtner et al., 2017). All metadata are published under the CC-0 license (https://creativecommons.org/).

The right of intellectual property remains unaffected at all times. It is specified in the metadata and is thus inseparably linked to its research data and can be viewed by the interested party at any time during reuse. This metadata entry is a prerequisite for the provision of the research data.

The BonaRes Data Centre fully supports the view that data should be freely available at all times without having to pass through a gatekeeper (FAIR: accessible). However, in some cases, the concerns of the data owner must be taken into account and thus the immediate provision of LTE data may be excluded, e.g., to ensure the right of first use. If a complete data release should not take place instantly, the BonaRes Data Centre developed a number of possibilities to be able to flexibly adapt the data policy, together with the data owner, to their particular requirements.

One example of how the right to first use for the data owner is implemented by the BonaRes Data Centre is the floating embargo. In this case, a complete dataset is transferred, which is regularly updated with additional values. As an example, yield data are assumed here, which are collected annually. The requirement of the data owner is that the current 5 years are not made publicly available. The solution of the BonaRes Data Centre is therefore to embargo these years. Floating now means that each year, when yield data are updated (T+1), they automatically fall into the embargo. At the same time, the oldest year affected by an embargo (T-4) is unlocked, compare Fig. 9.2.

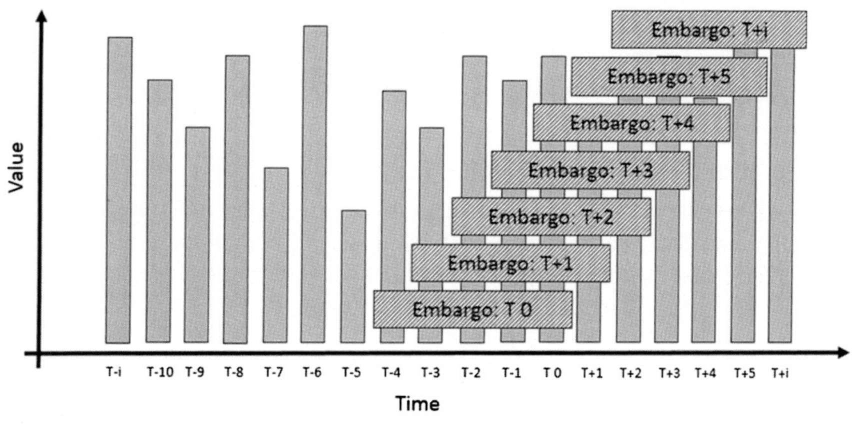

FIGURE 9.2

Schematic representation of the floating embargo as it can be applied to the publication of data in the Data Centre. The Y-axis indicates data values (e.g., yield of arable crops). The X-axis displays the time (e.g., harvest year). "T0" is the recent year. Gray bars represent the respective values. The shaded areas symbolize the embargo that is in effect at that particular time.

After the embargo has expired, data are provided with a Creative Commons license (CC) and a Digital Object Identifier (DOI), i.e., a permanent identification code to the research data, and are made available for free reuse via the BonaRes Data Portal (https://maps.bonares.de/mapapps/resources/apps/bonares).

Technical infrastructure

The technical infrastructure "BonaRes Repository" was set up to improve the access on geospatial soil-agricultural research data in Germany with special focus on the provision of research data from LTEs. It consists of various components such as the metadata editor, implementing a newly developed metadata schema, and the BonaRes Data Portal. The infrastructure was set up based on international standards, e.g., for metadata to (a) become part of international and national data infrastructure networks implementing the INSPIRE directive (The European Parliament and the Council of the European Union, 2007) and (b) make data citable by the registration of research datasets with a DOI, which requires the support of the DataCite Metadata Schema. The INSPIRE metadata schema, based on ISO 19115, is used to describe geospatial data to harmonize data exchange between spatial data infrastructures (SDIs). DataCite aims at improving research data citations by assigning DOIs to research data, which is required for data citations and data publications. The provision of metadata is important to ensure a proper discovery, reuse, and analysis of the research data. The BonaRes Metadata Schema (Gärtner et al., 2017) was created to be compliant to both the INSPIRE and DataCite metadata schemas. It has been implemented in a web-based metadata editor for capturing descriptive information about the research data. The following sections introduce the main components of the technical infrastructure of the BonaRes Repository: the BonaRes Metadata Schema, the Metadata Editor, and the BonaRes Data Portal.

The BonaRes Metadata Schema

The BonaRes Metadata Schema combines metadata elements of the international geospatial metadata standard INSPIRE (European Commission Joint Research Centre, 2007) and the international metadata standard for research data of DataCite (DataCite Metadata Working Group, 2016). By supporting the INSPIRE metadata schema, the BonaRes Metadata Schema ensures interoperability with other INSPIRE compliant SDI via the OGC CSW (Open Geospatial Consortium Inc., 2007) standard. In addition, the BonaRes metadata model can be mapped to the DataCite metadata standard (Version 4.0), which enables the assignment of a DOI. This, in turn, increases the value of the LTE dataset because it can now be used for data citations in publications or for data publications by the data owner.

Furthermore, the metadata elements and their respective subelements can be used to interlink columns or attributes of different tables by describing relationships between different datasets, which in turn facilitates the reuse of complex datasets by other researchers. The current version of the BonaRes Metadata Schema consists of 42 elements, of which 22 are mandatory and 20 are optional elements. Detailed information about the metadata elements and their properties can be found in Gärtner et al. (2017) and Specka et al. (2019).

Metadata Editor

The BonaRes Metadata Editor[4] is a web-based application that manages the metadata to the stored LTE research data in the infrastructure. It facilitates the process of creating and publishing metadata in the BonaRes Repository. After their registration, researchers can enter metadata for their research data, save drafts of metadata locally, or create templates that can be later used for the creation of other metadata. The built-in user administration ensures access to the metadata, which is not yet published, only by authorized persons. A user-friendly user interface supports researchers in the process of creating and editing metadata (Fig. 9.3). When researchers finish the metadata editing, a built-in function in the editor first checks the metadata for conformance to the BonaRes Metadata Schema. If conformity is confirmed, a review performed by the team of the BonaRes Repository

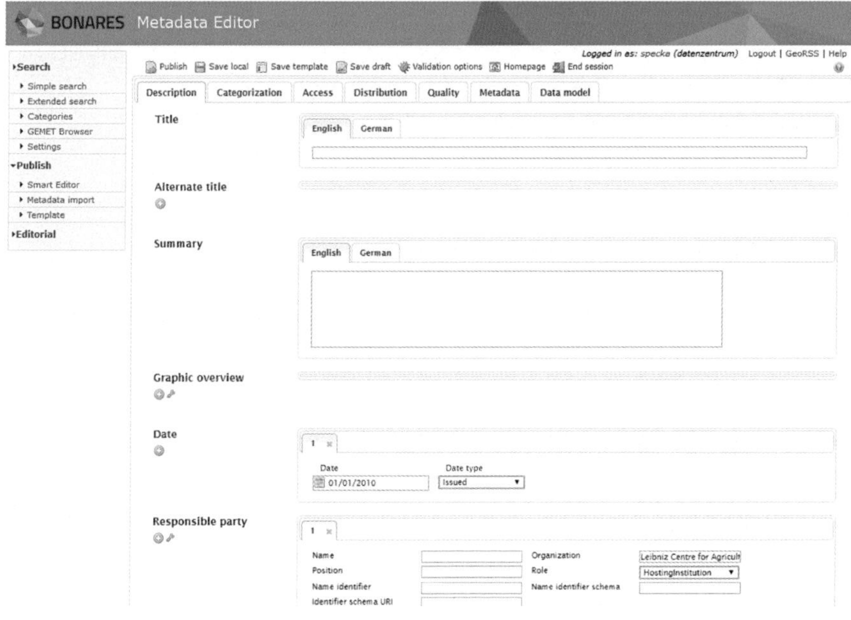

FIGURE 9.3

Metadata form for editing metadata in the BonaRes Metadata Editor.

4 https://metadata.bonares.de/terraCatalog/Start.do.

checks the metadata in terms of form and content. Only if metadata have passed this quality check, they will be made publicly available and accessible in the BonaRes Repository.

The BonaRes Data Portal

The BonaRes Data Portal[5] is a GIS-based web application (Fig. 9.4) and serves as the central access point for LTE data and other spatial soil-agricultural research data in Germany. It makes available information about research data managed within the BonaRes Repository as well as spatial data from external sources the Data Portal is interlinked with. Via the BonaRes Data Portal, metadata information from other INSPIRE compliant SDIs can be harvested and provided by supporting the INSPIRE metadata schema and implementing the OGC CSW interface (Open Geospatial Consortium Inc., 2007), which improves the dissemination of research data and, consequently, the discovery and findability of it. For example, the data portal is connected to the Product Center[6] of the Federal Institute for Geosciences and Natural Resources in Germany and the Edaphobase[7] (Burkhardt et al., 2014) of the Senckenberg Museum of Natural History Görlitz.

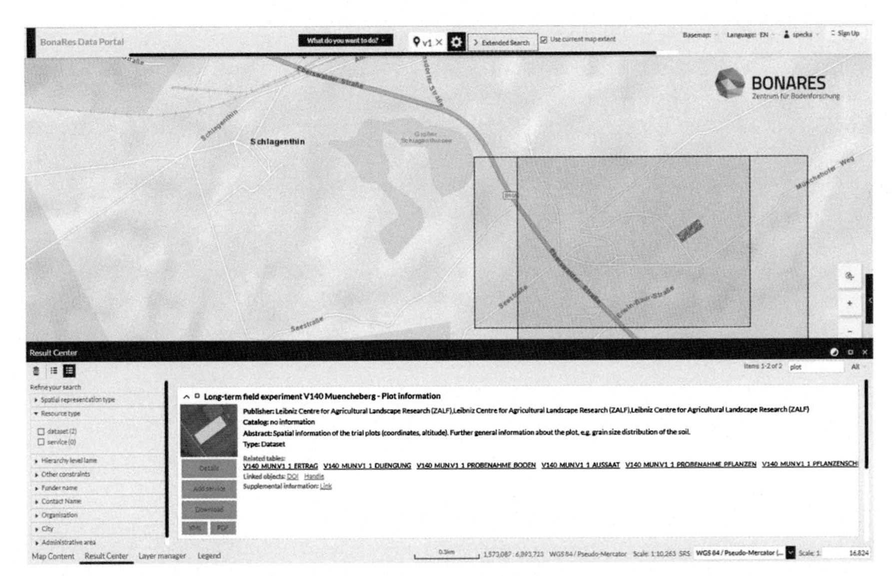

FIGURE 9.4

Presentation of search results for research data from the German LTE "V140 Müncheberg" and visualization of its plot design in the BonaRes Data Portal.

---

[5] https://maps.bonares.de/mapapps/resources/apps/bonares/index.html?lang=en.

[6] https://produktcenter.bgr.de/terraCatalog/Start.do.

[7] https://portal.edaphobase.org/.

Various search mechanisms have been implemented to improve the discovery of research data. The search is based on information from the metadata and search parameters can be specified by spatial, temporal, and thematic parameters. The built-in GIS allows the direct visualization of spatial geodata and standardized web services (e.g., OGC WMS) from internal or linked external sources in the Data Portal. Research data can be downloaded in common formats such as CSV, Excel-Files, or File-Geodatabase, a geodata format that integrates spatial and attribute information of datasets and that can be imported into various GIS applications.

Within the Data Portal, a DOI landing page is provided for each published dataset. Together with the provision of all metadata required by the DataCite metadata standard, the BonaRes Data Portal meets all requirements to assign a DOI for each dataset. Researchers can benefit significantly and increase their scientific output by creating data publications of datasets with a DOI stored in the BonaRes Data Portal. Thus they gain more credit for their work if their data are reused by others and accordingly cited via the DOI.

Publication of ZALF's LTE "V140"

One first example for an LTE research data publication is the LTE "V140" from Leibniz Centre for Agricultural Landscape Research (ZALF), Müncheberg, Germany. The V140 was established in 1963 at the experimental site of ZALF about 50 km east of Berlin in the district Märkisch-Oderland. It represents one of the few still active long-term field experiments on sandy soil. The most common soil subtype is Haplic Luvisol (FAO and IUSS, 2015). The experiment includes 21 treatments: five levels of mineral fertilization combined with four levels of organic fertilization and one control treatment. The dataset contains 32 parameters; among others: laboratory data of soil, plant samples, dates and application of sowing, fertilization and plant protection, plant varieties, and yield data. The data were quality checked in form and content, and methods were named and assigned to standards if possible.

By the publication of the dataset of the V140, the first LTE is now available via the BonaRes Portal for free reuse. The data can be downloaded in the BonaRes Repository via the BonaRes Data Portal (https://doi.org/10.20387/BonaRes-BSVY-R418). To ensure the initial scientific exploitation of the research data by the data owner (right of first use), the provision is delayed by a 9-year floating embargo. If more recent data are needed, the data owner can be contacted.

Conclusion

The BonaRes Repository offers simplified findability of suitable LTEs for the analysis of LTE data. Besides that, an easy and open access to the data stored in the BonaRes Repository is offered. Networking among different LTEs and the joint analysis of LTE data becomes easier through the provided information in the BonaRes Repository. Intensified analyses of LTE data will hopefully lead not only to a hub for agricultural and soil sciences but also promote valuable pieces of advice for practitioners and politicians.

The more frequent analyses of LTE data, at the same time, may make the LTE itself more valuable and may also help to maintain the LTE. Further benefits for the LTE, when data are provided for free via the BonaRes Repository, include the support in data digitalization (only possible for German LTE), processing, and storage (all LTEs). Furthermore, quality checks are conducted by the team of the BonaRes Repository.

Despite the tremendous scope and potential of the BonaRes repository, remaining challenges include the limited resources and capacities to digitize and transfer LTE metadata and data into the repository, the reservation among many LTE holders against sharing LTE data, and the limited resources to maintain LTEs themselves. However, the repository is meant to showcase the scientific value of LTE data which, in the long run, might safeguard the financial support to their maintenance.

Acknowledgments

This project is funded by the German Federal Ministry of Education and Research (BMBF) in the framework of the funding measure "Soil as a Sustainable Resource for the Bioeconomy—BonaRes", project "BonaRes (Module B): BonaRes Centre for Soil Research, subproject B" (grant 031B0511B).

References

Alliance of German Science Organisations, 2008. Priority Initiative "Digital Information", Berlin. Available from: http://www.dfg.de/download/pdf/foerderung/programme/lis/allianz_initiative_digital_information_en.pdf.

Amberger, A., Gutser, R., 1976. Effect of long-term potassium fertilization on crops and potassium dynamics of a brown earth. Annales Agronomiques 27, 643–657.

Berlin Declaration on Open Access to Knowledge in the Sciences and Humanities, 2003. Available from: https://openaccess.mpg.de/Berlin-Declaration.

BMBF (Federal Ministry of Education and Research), 2018. Research for a Biobased Economy. Success Stories and Challenges Facing the German Bioeconomy. Available from: https://www.bmbf.de/pub/Research_for_a_biobased_economy.pdf.

BMBF (Federal Ministry of Education and Research), BMEL (Federal Ministry of Food and Agriculture), 2014. Bioökonomie in Deutschland. Chancen für eine biobasierte und nachhaltige Zukunft. Available from: https://www.bundesregierung.de/breg-de/service/publikationen/biooekonomie-in-deutschland-736050.

Burkhardt, U., Russell, D.J., Decker, P., Döhler, M., Höfer, H., Lesch, S., Rick, S., Römbke, J., Trog, C., Vorwald, J., Wurst, E., Xylander, W.E.R., 2014. The Edaphobase project of GBIF-Germany – a new online soil-zoological data warehouse. Applied Soil Ecology 83, 3–12.

Debreczeni, K., Körschens, M., 2003. Long-term field experiments of the world. Archives of Agronomy and Soil Science 49 (5), 465–483.

DataCite Metadata Working Group, 2016. DataCite Metadata Schema Documentation for the Publication and Citation of Research Data. Version 4.0., DataCite e.V.

European Commission Joint Research Centre, 2007. INSPIRE Metadata Implementing Rules: Technical Guidelines based on EN ISO 19115 and EN ISO 19119.

FAO and IUSS, 2015. World Reference Base for Soil Resources 2014: International Soil Classification System for Naming Soils and Creating Legends for Soil Maps: Update 2015. World Soil Resources Reports No. 106. FAO, Rome.

Gärtner, P., Svoboda, N., Kühnert, T., Zoarder, M.M.A., Heinrich, U., 2017. The BonaRes Metadata Schema. BonaRes Centre. Available from: https://doi.org/10.20387/BonaRes-5PGG-8YRP

Grosse, M., Heinrich, U., Hierold, W., 2019. Fact Sheet for the Description of Long-Term Field Experiments/Steckbrief zur Beschreibung von Dauerfeldversuchen. BonaRes Centre. Available from: https://doi.org/10.20387/BONARES-R56G-FGRW

Herbst, F., Merbach, W., 2015. Mitteldeutsche Düngungsforschungszentren unter besonderer Berücksichtigung der Strohdüngung, vol. 28. Mitteilungen Agrarwissenschaften, Berlin.

Hoffmann, C., Schulz, S., Eberhardt, E., Grosse, M., Russell, D.J., Kühnert, T., Stein, S., Zoarder, M.A.M., Specka, X., Svoboda, N., Heinrich, U., 2018. Overview of Relevant Standards for the BonaRes-Program. Available from: https://doi.org/10.20387/BonaRes-9D25-0D93.

ISO 16269-4:2010, Statistical Interpretation of Data – Part 4: Detection and Treatment of Outliers.

ISO 19157:2013, Geographic Information – Data Quality.

ISO 28258:2013, Soil Quality – Digital Exchange of Soil-Related Data.

ISO/DIS 18400-100:2017, Soil Quality – Sampling- Part 100: Guidance on the Selection of Sampling Standards.

Körschens, M., 1990. Dauerfeldversuche: Übersicht, Entwicklung und Ergebnisse von Feldversuchen mit mehr als 20 Jahren Versuchsdauer, Akademie der Landwirtschaftswissenschaften.

Körschens, M., 1994. Der Statische Düngungsversuch Bad Lauchstädt nach 90 Jahren: Einfluss der Düngung auf Boden, Pflanze und Umwelt: mit einem Verzeichnis von 240 Dauerfeldversuchen der Welt. B.G. Teubner Verlagsgesellschaft.

Körschens, M., 1997. Die wichtigsten Dauerversuche der Welt - Übersicht, Bedeutung, Ergebnisse. Archiv für Acker- und Pflanzenbau und Bodenkunde 42, 157–168.

Körschens, M., 2000. Modellversuch zur Stalldungsteigerung Bad Lauchstädt. IOSDV: Internationale organische Stickstoffdauerdüngungsversuche. Bericht der Internationalen Arbeitsgemeinschaft Bodenfruchtbarkeit in der Internationalen Bodenkundlichen Union (IUSS). Bad Lauchstädt 15/2000, 133–134.

Open Geospatial Consortium Inc, 2007. OpenGIS® Catalogue Services Specification 2.0.2 – ISO Metadata Application Profile.

Specka, X., Gärtner, P., Hoffmann, C., Svoboda, N., Stecker, M., Einspanier, U., Senkler, K., Zoarder, M.A.M., Heinrich, U., 2019. The BonaRes metadata schema for geospatial soil-agricultural research data – Merging INSPIRE and DataCite metadata schemes. Computer & Geosciences 132, 33–41. https://doi.org/10.1016/j.cageo.2019.07.005.

Svoboda, N., Heinrich, U., 2017. The BonaRes Data Guideline. BonaRes Data Centre. Available from: https://doi.org/10.20387/BonaRes-E1AZ-ETD7

The European Parliament and the Council of the European Union, 2007. Directive 2007/2/EC of the European Parliament and of the Council of 14 March 2007 Establishing an Infrastructure for Spatial Information in the European Community (INSPIRE), 50, Official Journal of the European Union.

Wilkinson, M.D., Dumontier, M., Aalbersberg, I.J., Appleton, G., Axton, M., Baak, A., Blomberg, N., Boiten, J.-W., da Silva Santos, L.B., Bourne, P.E., Bouwman, J., Brookes, A.J., Clark, T., Crosas, M., Dillo, I., Dumon, O., Edmunds, S., Evelo, C.T., Finkers, R., Gonzalez-Beltran, A., Gray, A.J.G., Groth, P., Goble, C., Grethe, J.S., Heringa, J.,'t Hoen, P.A.C., Hooft, R., Kuhn, T., Kok, R., Kok, J., Lusher, S.J., Martone, M.E., Mons, A., Packer, A.L., Persson, B., Rocca-Serra, P., Roos, M., van Schaik, R., Sansone, S.-A., Schultes, E., Sengstag, T., Slater, T., Strawn, G., Swertz, M.A., Thompson, M., van der Lei, J., van Mulligen, E., Velterop, J., Waagmeester, A., Wittenburg, P., Wolstencroft, K., Zhao, J., Mons, B., 2016. Comment: the FAIR Guiding Principles for scientific data management and stewardship. Scientific Data 3.

CHAPTER

10 Long-term experiments on agroecology and organic farming: the Italian long-term experiment network

Corrado Ciaccia[1], Danilo Ceccarelli[2], Daniele Antichi[3], Stefano Canali[1]

[1]*CREA-Research Centre Agriculture & environment, Rome, Italy;* [2]*CREA-Research Centre for Olive, Citrus and Tree Fruit, Rome, Italy;* [3]*Department of Agriculture, Food and Environment, University of Pisa, Pisa, Italy*

Organic farming research agenda

It is now established that the focus on increasing yield at any cost promoted by the Green Revolution is not sustainable and is not sufficient to eradicate hunger and poverty to face the challenges of natural resources exhaustion, environment degradation, and biodiversity loss (FAO, 2017). In fact, current intensive and specialized food and farming systems are considered unable and insufficient to guarantee adequate and fair distribution of incomes and to mitigate and/or to adapt to climate change (IPES-Food, 2016). To achieve the 2030 Agenda for Sustainable Development (United Nations, 2018), there is an urgent need to promote transformative changes in how food is grown, produced, processed, transported, distributed, and consumed.

In its vision document about research and innovation for Sustainable Food and Farming, the European Technology Platform TP Organics (2017) ambitiously claimed that the transformation of the European food and farming systems toward sustainability by 2030 should imply the shift to at least 50% agroecological and organic agriculture, thus guaranteeing the implementation of the organic and agroecological principles as a source of inspiration for interdisciplinary research, for innovation on farms and in the food industry, and for building strong relations with farmers, consumers, and citizens.

This transformation trajectory is grounded on the application of three fundamental features, which should characterize the innovative farming and food systems, namely (i) efficiency (i.e., producing more food with less waste, less environmental pollution, and less land degradation), (ii) consistency (i.e., adapting production/transformation systems to their specific territorial, cultural, and socioeconomic context to be sustainable to let production and consumption be compatible with the carrying capacity of the ecosystems affected by it), and (iii) sufficiency (i.e., controlling increases in resources consumption, the so-called rebound effect).

Long-Term Farming Systems Research. https://doi.org/10.1016/B978-0-12-818186-7.00011-4
Copyright © 2020 Elsevier Inc. All rights reserved.

Turning agroecological principles into organic farming

Organic agriculture is faced with the challenge of feeding an increasing world population, ensuring simultaneously food security and environmental sustainability. Despite this, less than 1% of global farmland is managed organically and the global consumers of organic products are still very few (Willer and Lernoud, 2017). Moreover, the widespread development of (or "input substitution") organic production systems—characterized by the substitution of synthetic inputs with those allowed by the organic regulation and the mimic of the conventional upstream and downstream food supply chain organization—threatens organic farming and may affect the consumer trust on the sustainability of organic production (Darnhofer et al., 2010; Goldberger, 2011; EGTOP, 2013). These considerations highlight the need of taking organic out of its current niche and positioning it into the mainstream with the contemporary evolution of the so-called *Organic 3.0*, based on scientific evidences, reducing the trend to become globally standardized and uniquely business-oriented (Arbenz et al., 2015; Rahmann et al., 2017). In this framework, agroecology can be incorporated into organic research, allowing a holistic and multidisciplinary approach and facing the global challenges of the food system (Niggli, 2015; Ciaccia et al., 2019).

Agroecological approach

The function of organic cropping systems is based on biological/ecological processes and is characterized by similarities to natural ecosystems. Agroecosystems are overwhelmingly complex. While conventional agriculture has always promoted practices to reduce such complexity by using external synthetic chemical and energy inputs, organic agriculture is featured by the ability to sustain its farming practices with ecosystems functions embedded in that complexity.

Within the European Union (EU), organic farming is a certified production system regulated by law (i.e., Reg. CE 2007/834 and Reg. CE 2008/859 as modified by the newest Reg. CE 2018/848). Nevertheless, besides the compliance with regulations, what really qualifies organic systems, consistent with the general principles stated by IFOAM, is the agroecological approach. In some European countries, agroecology is described as "(1) the integrative study of entire food systems, encompassing ecological, economic and social dimensions, and (2) the design of individual farms using principles of ecology involving landscape, community and bioregion with emphasis on uniqueness of place and the people and other species that inhabit that place" (Francis et al., 2003). In this form, agroecology is an approach to the study and implementation of sustainable agricultural systems (i.e., not only the organic ones), basically applying three basic ecological properties to agroecosystems (Pacini and Groot, 2017):

- Diversity, which is given by the number of different components and processes present in the system and their relative abundance;

- Coherence, which provides measures of the numbers and strengths of the connections and flows among components and processes within the system;
- Connectedness, which is similar to coherence, but concerns the connections with components outside the agroecosystem.

Understanding and managing such systems requires long-term monitoring and assessment of biodiversity-related aspects of cropping systems. The concept of *functional biodiversity* (i.e., the part of the biosphere providing the desired ecosystem services) clearly describes the importance of agrobiodiversity in regulating the functioning of those cropping systems relying mostly on internal natural resources, as the organic ones are (Moonen and Bàrberi, 2008). Functional biodiversity is then at the core of designing, assessing, and improving organic cropping systems targeted to agroecological intensification.

From farm to food system

The goal of sustainably improving agricultural productivity to meet growing demand is agriculture's first challenge (FAO, 2019) and can be achieved by redesigning agrifood systems, moving from the globally standardized and business-oriented approach, encompassing the ecological, economic, and social dimensions of agriculture (Wezel et al., 2009; Gliessman, 2016). The system redesign approach is essentially based on the agroecological articulation of the production systems at cropping system and/or at farm scale, which, being more diversified in time and space, are able to supply a wide range of products in quantity and quality. These systems are characterized by diversified rotation, the implementation of intercropping, and agroecological service crops (e.g., cover crops, green manures, and break crops) and, in the most diversified and integrated models, by the combination of the plant, the animal, and the tree components (i.e., mixed farming, agroforestry, and agro-zoo-forestry systems).

This articulation generates downstream a range of supply chains and/or diversified food network; it implies multiple modes of interaction between the actors playing in their implementation and participating to the fair and responsible distribution of the product value (Meynard et al., 2013). Often, processors, retailers, and even consumers might aim to participate and to play an active role in the supply chain/network governance, contributing to guarantee the linkage among food productions, the territories where these activities are carried out, the local cultural values, and heritages.

The transition resulting from the system redesign and the consequent horizontal and vertical diversification of production systems to a wider spatial scale can determine further advantages of an ecological, economic, and social nature. Indeed, ecological functions and regulatory services can operate at scales larger than farm and/or cropping system and, when this happens, the design of agroecosystems must take into consideration the wider territorial context. Obviously, the transition from the farm to the territory scale implies a radical modification of decision-making and governance processes, which from individual becomes collective and which must, therefore, be based on fully implemented and effective participation mechanisms. For this reason, the shift toward an agroecology-based governance of territories and the creation of sustainable food systems require simultaneous technological and institutional innovations.

Long-term experiment as tool for research in organic farming

As aforementioned, the agroecological transition of agrifood systems should pass through redesign processes, to be realized at several levels, from the field to the food system scale. This implies the implementation of diversification strategies of cropping systems on space and time, as well as rethinking also the role of different actors of the agricultural sector through transdisciplinary, participatory, and change-oriented research and action activities. Long-term experiments (LTEs), therefore, might play a crucial role promoting transformation pathways of agrifood systems based on redesign, as they allow the long-term monitoring and assessment of the complex biological and ecological processes on which cropping system functioning is based on. Moreover, they should aim to tackle the research priorities for local producers, representing a potential hub of common knowledge and innovation. LTEs' setup and activities should hence be promoted to extract main information about the interconnection of the biological processes and the physical environment/the human management and contemporary test the research priorities identified together with different stakeholders.

Activities for process

Conventional research approach, in which activities are mainly related to the specific aims of a project, should be overcome to reach an "activities for process" approach, in which different projects differently contribute to the same long-term process, starting from its activation up to context analysis and processes evaluations. Indeed, research is often practised over short timescales and threatened by funding droughts and changes in stewardship (Owens, 2013). In agricultural research, exploring the effect of innovative practices (e.g., agroecological practices aimed at improving organic farming sustainability) on slow evolving parameters requires the support of experiments repeated in the long run (Rahmann et al. 2017). The change of perspective can drive into relevant results, strongly contributing to the agricultural system sustainability, as the evaluation of climate change and mitigation strategies, as well as the long-term effects of contrasting management options on biodiversity, and the related ecological services/disservices. The LTEs can then be considered as the training ground on which to apply this approach for knowledge and science implementation.

Participatory research

As part of a long-term process, multistakeholder involvement should be encouraged as it is acknowledged for boosting dynamic innovation (Delate et al., 2016), addressing the social, environmental, and economic sustainability goals of different interest groups (Van de Fliert and Braun, 2002). The involvement of local stakeholders in LTE activities is a feasible way to transfer innovation at the territorial scale through word of mouth and the organization of workshops and open day visits. Indeed, participatory research activities, rather than "collaborative" ones, allow to overcome the lock-in of single field management, reaching a collective point of view and goals likely different from the ones of participant themselves (Bruges and Smith, 2008). Stakeholders (e.g., farmers, consumers, processors, policymakers, and researchers) should collaborate in planning, management, and coordination of the LTEs, hence codesigning the cropping system represented by the same LTEs (Ciaccia et al., 2019). By this, once the benchmark of participation is decided, the LTE can become a territorial hub of innovation, in which (i) research demands derive directly from stakeholders, (ii) research issues are addressed and tested in the LTE, and (iii) results are discussed within the same stakeholder platform. The connection with satellite pilot farms should also be promoted to maximize the impact of the activities at local/territorial scale. The networking of experiences structured in this way can be considered a scaling-up strategy, answering to the main critics to participatory research projects of remaining isolated "islands of success" (El-Swaify and Evans, 1999, p. 37).

The Italian network of organic long-term experiments

From the local to the national scale

The experience on organic LTEs in Italy dates to 1991, when the MOLTE (Montepaldi Long Term Experiment) was set up as a comparison between organic and conventional arable systems in Tuscany (Migliorini and Vazzana, 2007). After MOLTE, several LTEs were set up in a wide range of pedoclimatic areas of the country and considering different production systems (i.e., fruit, arable, and vegetable systems) and different objectives, thus reflecting the multifaced structure of Italian agrifood system and the related needs in terms of innovation at local scale. Nowadays, LTEs' main research topics focus on crop diversification, fertilization strategies and waste recycling, genotype evaluation, produce traceability and quality, agroecological practices for weed, pest and disease management, sustainable production, and increased efficiency of utilization of water and available resources. Organic LTEs, therefore, are currently designed as multidisciplinary research facilities for testing holistic cropping/farming systems, which aim at supporting favorable ecological and biological processes to improve agroecosystem functioning. To date, LTEs are also used as field laboratories in participatory research and in the activity of checking innovations, allowing the continuous comparison and exchange of information among the operators of the organic value chains and the national and international scientific communities. Their functionality and efficiency are basilar tools to guarantee the effectiveness of the dissemination and transfer of research innovations to farmers and other stakeholders.

With the final goal of further enhancing the impact of the Italian LTEs, the RetiBio project, funded by the Italian Food and Farming Office of the Ministry of Agriculture, was launched in 2014. The project had the main objective of supporting the operational potential of the individual LTE and creating a network among scientists involved in organic food and farming research based on the organic LTEs (Peronti et al., 2015). Harmonization of research approaches and expansion of the opportunity for collaborative research have been the most relevant outcomes of the RetiBio experience that, thanks to its acknowledged success, was extended in the following RetiBio 2 project. The network endorsed by the RetiBio and RetiBio 2 projects encompasses organic LTEs either managed by research teams of the CREA, the Italian largest research institution for the agrifood sector, or of a number of Italian Universities (Table 10.1, Fig. 10.2). The realized interinstitutional collaborative research activities contributed to activate a lively debate, in the frame of which the need for an international dimension of the LTE network was conceptualized, paving the road to the following European networking initiatives (Fig. 10.1).

Next steps

The new project PERILBIO (Promotion and strengthening of LTEs in organic farming), funded in 2019 by the Italian Ministry of Agricultural, Food and Forestry Policies (MIPAAFT), will ensure the maintenance, strengthening, and enhancing of CREA LTEs network in the next future; in addition to ensuring the continuation of ongoing activities, this program involves the set up of new LTEs. In particular, PERILBIO will be focused on the following: (i) realization of three new LTEs for animal production (poultry farming, cuniculture, and aquaculture in organic systems); (ii) consolidation and expansion of the relationship among CREA LTEs network and operators of the organic sector to improve innovation transfer; (iii) implementation of the participatory approach and development of research and experimentation activities already carried out in previous initiatives; and (iv) creation of new biodistricts, in organic production areas, which can benefit from LTE research to test activities and promote outreach.

European long-term experiment network

Although LTEs are nowadays considered as patrimonial research facilities by many institutions, the financial support of these experiments is still very scattered, discontinuous, and of moderate importance, being conveyed by funding for single research projects lasting for no more than 3–4 years each. This context generated huge

Table 10.1 RetiBio 2 network of Italian organic long-term experiment (LTE).

LTE acronym	LTE complete name	Institution in charge of the LTE and site	Setup year	Cropping system	Characteristics	Additional information	References
BIO-CONV	Long-term effects of conventional and organic cropping systems	Tuscia University (Viterbo— Central Italy)	2003	Arable	Conventional versus organic rotation × plowing versus subsoiling	Evaluation of the long-term dynamics of soil fertility parameters, crop yields, system biodiversity (weeds, pests, and diseases)	Campiglia et al. (2015)
BIOLEA	Long-term organic table olive experiment	CREA—Research Centre for Olive, Citrus and Tree Fruit Acireale (Sicily)	2015	Olive	Cultivar × management comparison	Compared systems combine organic amendments, green manure, and different irrigation strategies	Peronti et al. (2015)
BIOSYST	N/a	University of Perugia (Central Italy)	1971 Organic since 2002	Arable/ vegetable	Low input versus organic rotation	Assessing the long-term average yield level, stability, and sustainability of compared cropping systems	Bonciarelli et al. (2016)
MAIOR	Maintenance of organic orchards	CREA—Research Centre for Olive, Citrus and Tree Fruit Rome (Central Italy)	2010	Stone fruit	Cultivar × management comparison	Comparison of input substitution versus agroecological systems based on diversification. Participatory activities	Peronti et al. (2015)

Continued

Table 10.1 RetiBio 2 network of Italian organic long-term experiment (LTE).—*cont'd*

LTE acronym	LTE complete name	Institution in charge of the LTE and site	Setup year	Cropping system	Characteristics	Additional information	References
MASCOT	Mediterranean arable system comparison trial	University of Pisa Centre for Agri-Environmental Research "Enrico Avanzi" (CiRAA) Scuola Superiore Sant'Anna (SSSA) San Piero a Grado, Pisa (Tuscany)	2001	Arable	Conventional versus organic rotation	Evaluation of the long-term dynamics of soil fertility parameters, system biodiversity, as well as agronomic, economic, and energetic aspects Best practices and agronomic solutions are tested in the compared systems	Bàrberi and Mazzoncini (2006)
MITIORG	Long-term climatic change adaptation in organic farming	CREA—Research Centre for Agriculture and Environment Metaponto (South of Italy)	2014	Vegetable	Climate change mitigation strategies	Agronomic practices and diversification practices to reduce the effect of extreme events in a semiarid area	Diacono et al. (2016)
MOLTE	Montepaldi Long-Term Experiment	University of Florence (Tuscany)	1991	Arable	Conventional versus organic rotation	Implementation of agroecological practices in organic rotation and comparison of cropping systems	Migliorini and Vazzana (2007)

MOVE LTE	Monsampolo Vegetable organic long-term field experiment	CREA—Research Centre for Agricultural and Environment Monsampolo del Tronto (Central Italy)	2001	Vegetable	Diversification and agronomic strategies	Agroecosystem evaluation in terms of agronomic, economic, and environmental sustainability, with particular regard to the quality of the fresh and transformed production. Participatory activities	Campanelli and Canali (2012)
PALAP9	Long-term trial on organic citrus	CREA—Research Centre for Olive, Citrus and Tree Fruit Acireale (Sicily)	1995	Citrus	System management comparison	Evaluation of the effect of citrus production wastes, by-products, and biomass of animal origin use as fertilizers on fruit quality, plant nutritional status, and soil fertility status	Canali et al. (2009)

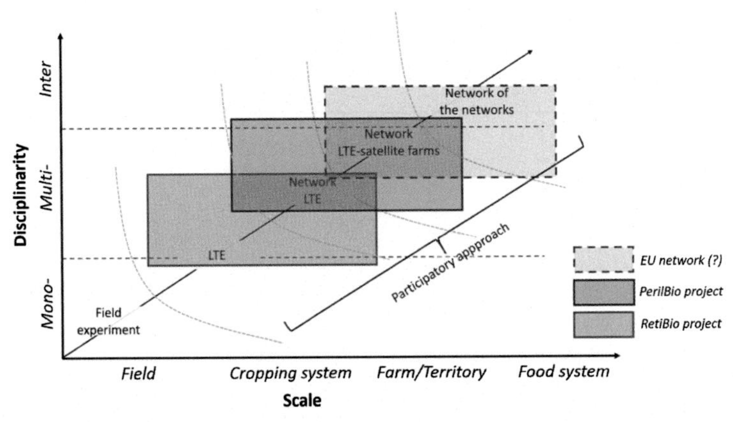

FIGURE 10.1

Pathway of networking of the Italian long-term experiments (LTEs) and implication on OFF research approaches.

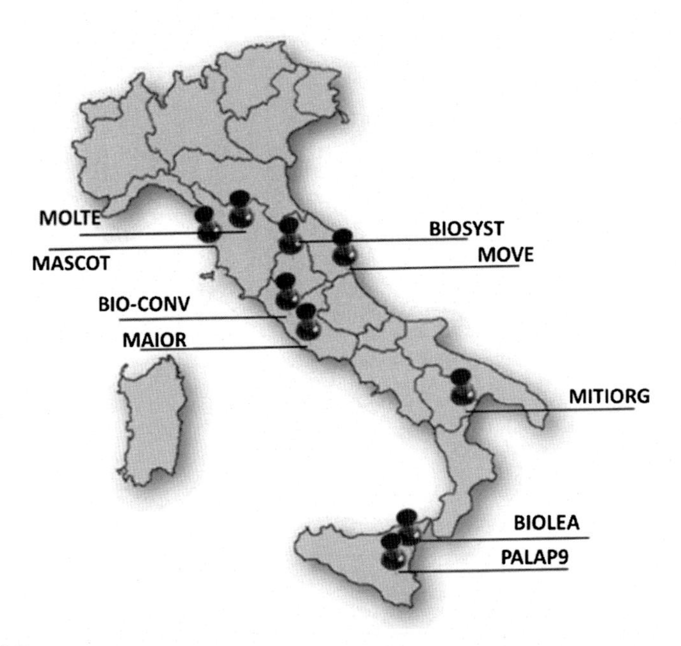

FIGURE 10.2

The Italian long-term experiment network (2018).

differences in the objectives, management, and duration among all the organic LTEs developed in Europe, which in the end reduced the potential impact of research in organic farming—a potential that could be explored by networking the single organic LTEs. To the best of our knowledge, the first organic LTEs networking experience was initiated in 2006 within the International Society of Organic Agriculture Research (ISOFAR) with the establishment of a working group on organic farming LTEs (Raupp et al., 2006). In 2008, the working group organized a scientific session in the frame of the Second Scientific Conference of ISOFAR, held at the 16th IFOAM Organic World Congress in Modena, Italy. Subsequently, the Working Group was not kept active anymore even if some networks arose and grew on a national scale thereafter.

Recently, two national networks of organic LTEs were identified and linked: the already mentioned RetiBio network in Italy and RotAB network in France. They organized cross-visits and workshops (one at the Agroecology Europe Forum in Lyon, 2017; one at the GRAB-IT International Workshop held in Capri, 2018) and have been sharing experiences on LTE's management, fund raising opportunities, and stakeholder involvement. From these first meetings emerged the idea to structure a European network of organic LTEs, thus revitalizing the former experience promoted by ISOFAR some year ago and fostering—at the same time—the discussion dealing with the organic LTEs to put it in line with the most updated research trajectories of the organic food and farming sector. Indeed, it was considered relevant to include in the network organic LTEs that are designed in accordance with the agroecological principles, then distinguishable for their nature and features. The establishment of such a network will be aimed at (i) reducing knowledge gaps and harmonizing research approaches, for a higher impact of the results; (ii) promoting the quality and quantity of research for the transition of the agrifood systems to become more sustainable; (iii) expanding the opportunity for collaborative, multiactor research and coinnovation actions across the community and even outside (e.g., with scientists active on organic farming and agroecology even outside the EU); and (iv) paving the road for further initiatives in the area of European agrifood system research infrastructure development.

Future perspectives

Although the above-described experiments are suitable to carry out researches in cropping systems that are economically relevant for Italian organic agriculture, to date, LTEs representative of some of the most typical organic productions of Italy are not included and/or underrepresented in the network (i.e., vineyard, rice, agroforestry, mixed farming). Therefore, in addition to supporting actions addressed to maintain the existing experiments in the long term, future investments to reinforce and to enrich the Italian LTEs network should be foreseen, by strengthening this objective in the Italian and in the European organic food and farming research agenda.

The outcomes obtained so far have also allowed establishing criteria and principles which should be followed by research managers and policymakers to design effective research actions to enhance the impact of the LTEs and their networks. Indeed, five basic features have been identified, namely

- to guarantee, as minimum standard, the multidisciplinarity of research activities run in LTEs and to promote interdisciplinarity;
- to ensure local stakeholder involvement in defining, updating, and/or upgrading of LTEs' research objectives, as well as to promote an active engagement of actors in the definition of the management rules as well as results assessment and interpretation;
- to design and to pursue multiscale research actions, addressing issues at cropping system, farm, and supply chain scale when necessary;
- to promote LTEs' national and international networking activities aimed to share and to harmonize methodologies;
- to facilitate interinstitutional research collaborations to exploit synergies among Universities, Governmental and local/regional research and innovation institutions (either public or private) facilitating the integration of the scientific communities and enhancing knowledge creation and transferring.

References

Arbenz, M., Gould, D., Stopes, C., 2015. Organic 3.0 for truly sustainable farming and consumption. In: Discussion Papers Based on Think Tanking by SOAAN & IFOAM Organics International and Launched at the ISOFAR International Organic EXPO.

Bàrberi, P., Mazzoncini, M., 2006. The MASCOT (mediterranean arable systems comparison trial) long-term experiment (Pisa, Italy). Long-term field experiments in organic farming. ISOFAR. Science Series (1), 1−15.

Bonciarelli, U., Onofri, A., Benincasa, P., Farneselli, M., Guiducci, M., Pannacci, E., Tosti, G., Tei, F., 2016. Long-term evaluation of productivity, stability and sustainability for cropping systems in Mediterranean rainfed conditions. European Journal of Agronomy 77, 146−155.

Bruges, M., Smith, W., 2008. Participatory approaches for sustainable agriculture: a contradiction in terms? Agriculture and Human Values 25 (1), 13−23.

Campanelli, G., Canali, S., 2012. Crop production and environmental effects in conventional and organic vegetable farming systems: the case of a long-term experiment in Mediterranean conditions (Central Italy). Journal of Sustainable Agriculture 36 (6), 599−619.

Campiglia, E., Mancinelli, R., De Stefanis, E., Pucciarmati, S., Radicetti, E., 2015. The long-term effects of conventional and organic cropping systems, tillage managements and weather conditions on yield and grain quality of durum wheat (Triticum durum Desf.) in the Mediterranean environment of Central Italy. Field Crops Research 176, 34−44.

Canali, S., Di Bartolomeo, E., Trinchera, A., Nisini, L., Tittarelli, F., Intrigliolo, F., Roccuzzo, G., Calabretta, M.L., 2009. Effect of different management strategies on soil quality of citrus orchards in Southern Italy. Soil Use and Management 25 (1), 34−42.

Ciaccia, C., Di Pierro, M., Testani, E., Roccuzzo, G., Cutuli, M., Ceccarelli, D., 2019. Participatory research towards food system redesign: Italian case study and perspectives. Sustainability 11 (24), 7138.

Ciaccia, C., Testani, E., Roccuzzo, G., Canali, S., 2019. The role of agrobiodiversity in sustainable food systems design and management. In: Genetic Diversity in Horticultural Plants. Springer, Cham, pp. 245−271.

Darnhofer, I., Lindenthal, T., Bartel-Kratochvil, R., Zollitsch, W., 2010. Conventionalisation of organic farming practices: from structural criteria towards an assessment based on organic principles. A Review Agronomy for Sustainable Development 30, 67–81.

Delate, K., Canali, S., Turnbull, R., Tan, R., Colombo, L., 2016. Participatory organic research in the USA and Italy: across a continuum of farmer-researcher partnerships. Renewable Agriculture and Food Systems 32 (4), 331–348.

Diacono, M., Fiore, A., Farina, R., Canali, S., Di Bene, C., Testani, E., Montemurro, F., 2016. Combined agro-ecological strategies for adaptation of organic horticultural systems to climate change in Mediterranean environment. Italian Journal of Agronomy 11 (730), 85–91.

EGTOP (Expert Group for Technical Advice on Organic Production), 2013. Final Report on Greenhouse Production (Protected Cropping). Available from: http://ec.europa.eu/agriculture/organic/eu-policy/expert-advice/documents/final-reports/final_report_egtop_on_greenhouse_production_en.pdf.

El-Swaify, S.A., Evans, D.O., 1999. Sustaining the global farm: strategic issues, principles, and approaches. In: Prepared for the 10th International Soil Conservation Organization Conference, Purdue University, USA.

FAO, 2017. The Future of Food and Agriculture—Trends and Challenges.

FAO, 2019. Scaling up agroecology to achieve the sustainable development goals. In: Proceedings of the Second FAO International Symposium. Rome, 412 pp. Licence: CC BY-NC-SA 3.0 IGO.

Francis, C., Lieblein, G., Gliessman, S., Breland, T.A., Creamer, N., Harwood, R., Salomonsson, L., Helenius, J., Rickerl, D., Salvador, R., Wiedenhoeft, M., Simmons, S., Allen, P., Altieri, M., Flora, C., Poincelot, R., 2003. Agroecology: the ecology of food systems. Journal of Sustainable Agriculture 22 (3), 99–118. Available from: https://doi.org/10.1300/J064v22n03_10.

Gliessman, S., 2016. Transforming food systems with agroecology. Agroecology and Sustainable Food Systems 40, 187–189.

Goldberger, J.R., 2011. Conventionalization, civic engagement, and the sustainability of organic agriculture. Journal of Rural Studies 27, 288–296.

IPES-Food, 2016. From uniformity to diversity: a paradigm shift from industrial agriculture to diversified agroecological systems. In: International Panel of Experts on Sustainable Food Systems. www.ipes-food.org.

Meynard, J.M., Messéan, A., Charlier, A., Charrier, F., Farès, M., Le Bail, M., Magrini, M.B., Savini, I., 2013. Crop Diversification: Obstacles and Levers: Study of Farms and Supply Chains. Synopsis of the Study Report. INRA, p. 52.

Migliorini, P., Vazzana, C., 2007. Biodiversity indicators for sustainability evaluation of conventional and organic agro-ecosystems. Italian Journal of Agronomy 2 (2), 105–110.

Moonen, A.C., Bàrberi, P., 2008. Functional biodiversity: an agroecosystem approach. Agriculture, Ecosystems & Environment 127, 7–21.

Niggli, U., 2015. Incorporating agroecology into organic research-an ongoing challenge. Sustainable Agriculture Research 4 (3), 149–157.

Owens, B., 2013. Long-term research: slow science. Nature News 495 (7441), 300.

Pacini, G., Groot, J., 2017. Sustainability of agricultural management options under a systems perspective. In: Abraham, M. (Ed.), Encyclopedia of Sustainable Technologies. Elsevier, pp. 191–200.

Peronti, M., Bàrberi, P., Campanelli, G., Ceccarelli, D., Ceglie, F.G., Ferlito, F.S., Mazzoncini, M., Montemurro, F., Roccuzzo, G., Tittarelli, F., Riva, F., Ranuzzi, M.,

Canali, S., 2015. The Italian organic long term field experiments network. In: IFOAM Agribiomediterraneo International Conference "Agroecology for Organic Agriculture in the Mediterranean". 10-13 September 2015, Vignola Castle (Modena) and SANA Bologna, Italy.

Rahmann, G., Ardakani, M.R., Bàrberi, P., Boehm, H., Canali, S., Chander, M., Wahyudi, D., Dengel, L., Erisman, J.W., Galvis-Martinez, A.C., Hamm, U., Kahl, J., Köpke, U., Kühne, S., Lee, S.B., Løes, A.K., Moos, J.A., Neuhof, D., Tapani Nuutila, J., Olowe, V., Oppermann, R., Rembiałkowska, E., Riddle, J., Rasmussen, I.A., Shade, J., Mok Sohn, S., Tadesse, M., Tashi, S., Thatcher, A., Uddin, N., von Fragstein und Niemsdorff, P., Wibe, A., Wivstad, M., Wenliang, W., Zanoli, R., 2017. Organic Agriculture 3.0 is innovation with research. Organic Agriculture 7 (3), 169–197.

Raupp, J., Pekrun, C., Oltmanns, M., Köpke, U., 2006. Long-term Field Experiments in Organic Farming. Verlag Dr. H.J. Köster, Berlin.

TP Organics, 2017. Research & innovation for sustainable food and farming. In: Position Paper on the 9th EU Research & Innovation Framework Programme (FP9). https://tporganics.eu/. (Accessed April 2018).

United Nations, 2018. Agriculture Development, Food Security and Nutrition. Report of the Secretary-General. UN General Assembly. A/73/150.

Van de Fliert, E., Braun, A.R., 2002. Conceptualizing integrative, farmer participatory research for sustainable agriculture: from opportunities to impact. Agriculture and Human Values 19 (1), 25–38.

Wezel, A., Bellon, S., Doré, T., Francis, C., Vallod, D., David, C., 2009. Agroecology as a science, a movement and a practice. A review. Agronomy for sustainable development 29 (4), 503–515.

Willer, H., Lernoud, J., 2017. The World of Organic Agriculture. Statistics and Emerging Trends 2017 (pp. 1–336). Research Institute of Organic Agriculture FiBL and IFOAM-Organics International.

Index

'*Note*: Page numbers followed by "t" indicate tables and "f" indicate figures.'

A

B

C

D

E

F

G

H

I

J

Printed in the United States
By Bookmasters